Structural Dynamics of Macromolecules

Guest Editor

Ki Hyun Nam

MDPI

Basel • Beijing • Wuhan • Barcelona • Belgrade • Novi Sad • Cluj • Manchester

Guest Editor
Ki Hyun Nam
College of General Education
Kookmin University
Seoul
Republic of Korea

Editorial Office
MDPI AG
Grosspeteranlage 5
4052 Basel, Switzerland

This is a reprint of the Special Issue, published open access by the journal *International Journal of Molecular Sciences* (ISSN 1422-0067), freely accessible at: https://www.mdpi.com/journal/ijms/special_issues/FIFKQS1921.

For citation purposes, cite each article independently as indicated on the article page online and as indicated below:

Lastname, A.A.; Lastname, B.B. Article Title. *Journal Name* **Year**, *Volume Number*, Page Range.

ISBN 978-3-7258-8075-1 (Hbk)
ISBN 978-3-7258-8076-8 (PDF)
https://doi.org/10.3390/books978-3-7258-8076-8

Cover image courtesy of Ki Hyun Nam

Contents

About the Editor . . . . **vii**

Ki Hyun Nam
Special Issue "Structural Dynamics of Macromolecules"
Reprinted from: *Int. J. Mol. Sci.* **2026**, *27*, 961, https://doi.org/10.3390/ijms27020961 . . . . **1**

Dmitry Makarov, Anastasia Kharlamova
Peculiarities of Scattering of Ultrashort Laser Pulses on DNA and RNA Trinucleotides
Reprinted from: *Int. J. Mol. Sci.* **2022**, *23*, 15417, https://doi.org/10.3390/ijms232315417 . . . . . **4**

Tingting Bu, Jing Lan, Inseong Jo, Jie Zhang, Xue Bai, Shanru He, et al.
Structural Basis of the Inhibition of L-Methionine γ-Lyase from *Fusobacterium nucleatum*
Reprinted from: *Int. J. Mol. Sci.* **2023**, *24*, 1651, https://doi.org/10.3390/ijms24021651 . . . . . . **14**

Xue Bai, Jing Lan, Shanru He, Tingting Bu, Jie Zhang, Lulu Wang, et al.
Structural and Biochemical Analyses of the Butanol Dehydrogenase from *Fusobacterium nucleatum*
Reprinted from: *Int. J. Mol. Sci.* **2023**, *24*, 2994, https://doi.org/10.3390/ijms24032994 . . . . . . **29**

Jaewon Yang, Hyung Jin Jeon, Seonha Park, Junga Park, Seonhye Jang, Byeongmin Shin, et al.
Structural Insights and Catalytic Mechanism of 3-Hydroxybutyryl-CoA Dehydrogenase from *Faecalibacterium Prausnitzii* A2-165
Reprinted from: *Int. J. Mol. Sci.* **2024**, *25*, 10711, https://doi.org/10.3390/ijms251910711 . . . . . **45**

Mousheng Wu, Nusrat Jahan, Amanda Sharp, Anwar Ullah, Corinne E. Augelli-Szafran, Sixue Zhang and Rebecca J. Boohaker
Structure Characterization of Zinc Finger Motif 1 and 2 of GLI1 DNA Binding Region
Reprinted from: *Int. J. Mol. Sci.* **2024**, *25*, 13368, https://doi.org/10.3390/ijms252413368 . . . . . **61**

Han-ul Kim, Myeong Seon Jeong, Mi Young An, Yoon Ho Park, Sun Hee Park, Sang J. Chung, et al.
Comparative Analysis of Symmetry Parameters in the E2 Inner Core of the Pyruvate Dehydrogenase Complex
Reprinted from: *Int. J. Mol. Sci.* **2024**, *25*, 13731, https://doi.org/10.3390/ijms252413731 . . . . . **75**

Fernando Hinostroza, Sofía Albornoz-Muñoz, Sebastián Vergara, Gabriela Urra, Ingrid Araya-Durán, Rafael A. Fissore, et al.
Structural Implications of H233L and H398P Mutations in Phospholipase Cζ: A Full-Atom Molecular Dynamics Study on Infertility-Associated Dysfunctions
Reprinted from: *Int. J. Mol. Sci.* **2025**, *26*, 4706, https://doi.org/10.3390/ijms26104706 . . . . . . **85**

Sai-Nan Wang, Xiao-Xia Liu, Ling-Jie Lei, Qiang Wang, Zhu-Qing Shao and Yang Liu
Evolutionary Origins and Functional Diversification of 2′-O-Methyltransferases: Insights from Phylogenetic and Structural Analysis
Reprinted from: *Int. J. Mol. Sci.* **2025**, *26*, 5260, https://doi.org/10.3390/ijms26115260 . . . . . . **100**

About the Editor

Ki Hyun Nam

Ki Hyun Nam is a scientist at Kookmin University in the Republic of Korea. His research focuses on structural biology and macromolecular crystallography, with particular interest in enzyme structure and function. His work involves the determination and analysis of protein structures to understand molecular mechanisms at atomic resolution. He has extensive experience in X-ray crystallography, including serial crystallography and studies on the structural dynamics of macromolecules. His research also explores the application of computational approaches to analyze protein structures and molecular dynamics. Through his work, he aims to provide insights into enzyme mechanisms and to contribute to a broader understanding of biological macromolecules and their functions.

International Journal of
Molecular Sciences

MDPI

Editorial

Special Issue "Structural Dynamics of Macromolecules"

Ki Hyun Nam

College of General Education, Kookmin University, Seoul 02707, Republic of Korea; structure@kookmin.ac.kr

Understanding the molecular function of macromolecules is fundamental to understanding their biological roles [1,2]. In particular, macromolecular structures provide intuitive insights into their functions, including how biomolecules interact with substrates, ligands, and other biomolecules [3,4]. Various biotechnical techniques, such as X-ray crystallography, nuclear magnetic resonance (NMR), cryo-electron microscopy (cryo-EM), and micro-electron diffraction (microED), have been widely used to analyze high-resolution molecular structures [5–7] and provide insights into molecular flexibility and conformational states, while also being potentially useful for understanding the mechanisms of action of biological macromolecules [8–10]. In addition, time-resolved serial crystallography using X-ray free-electron lasers (XFELs) or synchrotron radiation, as well as time-resolved cryo-EM, provide structural insights into the dynamics of biomolecules during reaction processes [11–14]. Furthermore, molecular dynamics (MD) simulations can be used to obtain information on the motions and dynamics of macromolecules that are experimentally challenging to characterize [15]. These structural data not only offer valuable information for accurately understanding biological functions but also provide insights into protein engineering in the fields of drug discovery and industrial applications [16,17]. This Special Issue includes structural analyses of biological macromolecules, related technical and computational advancements, enzymatic mechanisms and industrial applications, and regulation and evolution, providing an important foundation for understanding macromolecular functions.

Makarov and Kharlamova demonstrate that pulse length is the dominant parameter governing the scattering of attosecond ultrashort pulses from DNA and RNA trinucleotides, leading to diffraction patterns that differ substantially from those predicted by conventional scattering theory. These findings suggest that time-resolved X-ray diffraction using attosecond pulses requires pulse-duration-dependent scattering models to correctly interpret and decode DNA and RNA structures.

Bu et al. reported the crystal structure of L-methionine γ-lyase (MGL) in *Fusobacterium nucleatum*, a lesion-associated obligate anaerobic pathogen implicated in destructive periodontal disease. Their structural analysis provides mechanistic insights into the molecular function of MGL and offers a structural basis for the rational design of potential inhibitors targeting this enzyme.

Bai et al. provide a biochemical and structural characterization of butanol dehydrogenase (BDH) from *Fusobacterium nucleatum*, highlighting its potential relevance for butanol biosynthesis. Their study identified substrate specificity, cofactor dependence, and key residues involved in substrate and cofactor binding, providing mechanistic insights into the molecular function of BDH and a structural basis for protein engineering toward industrial applications.

Yang et al. provided a structural characterization of 3-hydroxybutyryl-CoA dehydrogenase (A2HBD) from *Faecalibacterium prausnitzii* A2-165, a key enzyme involved in the butyrate production pathway associated with atopic dermatitis. They elucidated the cofactor and substrate binding modes, substrate-induced conformational changes, and key residues involved in catalysis, providing mechanistic insights into enzyme function and a structural basis for understanding butyrate biosynthesis.

Wu et al. reported on the novel homodimeric crystal structure of the N-terminal zinc finger motifs (Zif1–2) of the transcription factor GLI1, which is distinct from the DNA-bound conformation. Structural and computational analyses suggest that this dimeric arrangement may represent the native apo form of GLI1, providing new insights into its DNA-binding regulation.

Kim et al. demonstrated the effect of symmetry parameter selection in cryo-EM single-particle analysis by reconstructing the E2 inner core of the pyruvate dehydrogenase complex under multiple symmetry assumptions. The results show that near-atomic structures can appear identical despite different processing parameters, highlighting that improper parameter choices may obscure biologically meaningful dynamic information and require careful interpretation.

Hinostroza et al. utilized all-atom molecular dynamics simulations to investigate how infertility-associated mutations in sperm-specific phospholipase Cζ disrupt its function during fertilization. The results show that H233L and H398P mutations impair PIP2 and Ca^{2+} binding through altered interactions and substrate misalignment, whereas R553P mainly affects structural stability and dynamics, providing a molecular explanation for defective egg activation and male infertility.

Wang et al. systematically classified of ribose 2′-O-methyltransferases (2′-O-MTases) into 11 families and revealed their widespread distribution across the Tree of Life, with several families tracing back to the Last Universal Common Ancestor. Evolutionary and structural analyses showed that most 2′-O-MTases are conserved under purifying selection while undergoing lineage-specific expansion and functional diversification, particularly exemplified by the FtsJ family specializing in distinct RNA substrates.

A wide range of macromolecule-related studies are included in this Special Issue, addressing structural dynamics and evolutionary perspectives in addition to macromolecular structures. The contributions extend from fundamental science to applied research, and together they provide valuable insights that will facilitate a deeper understanding of the related fields.

Acknowledgments: I thank the contributing authors and the external reviewers for the opportunity to produce this Special Issue. During the manuscript preparation, the author used ChatGPT-5.2 for minor corrections and adjustments to the English language. The author has reviewed and edited the output and takes full responsibility for the content of this publication.

Conflicts of Interest: The author declares no conflicts of interest.

List of Contributions:

1. Makarov, D.; Kharlamova, A. Peculiarities of Scattering of Ultrashort Laser Pulses on DNA and RNA Trinucleotides. *Int. J. Mol. Sci.* **2022**, *23*, 15417. https://doi.org/10.3390/ijms232315417.
2. Bu, T.; Lan, J.; Jo, I.; Zhang, J.; Bai, X.; He, S.; Jin, X.; Wang, L.; Jin, Y.; Jin, X.; et al. Structural Basis of the Inhibition of L-Methionine γ-Lyase from *Fusobacterium nucleatum*. *Int. J. Mol. Sci.* **2023**, *24*, 1651. https://doi.org/10.3390/ijms24021651.
3. Bai, X.; Lan, J.; He, S.; Bu, T.; Zhang, J.; Wang, L.; Jin, X.; Mao, Y.; Guan, W.; Zhang, L.; et al. Structural and Biochemical Analyses of the Butanol Dehydrogenase from *Fusobacterium nucleatum*. *Int. J. Mol. Sci.* **2023**, *24*, 2994. https://doi.org/10.3390/ijms24032994.
4. Yang, J.; Jeon, H.J.; Park, S.; Park, J.; Jang, S.; Shin, B.; Bang, K.; Hawkes, H.-J.K.; Park, S.; Kim, S.; et al. Structural Insights and Catalytic Mechanism of 3-Hydroxybutyryl-CoA Dehydrogenase from *Faecalibacterium prausnitzii* A2-165. *Int. J. Mol. Sci.* **2024**, *25*, 10711. https://doi.org/10.3390/ijms251910711.

5. Wu, M.; Jahan, N.; Sharp, A.; Ullah, A.; Augelli-Szafran, C.E.; Zhang, S.; Boohaker, R.J. Structure Characterization of Zinc Finger Motif 1 and 2 of GLI1 DNA Binding Region. *Int. J. Mol. Sci.* **2024**, *25*, 13368. https://doi.org/10.3390/ijms252413368.
6. Kim, H.; Jeong, M.S.; An, M.Y.; Park, Y.H.; Park, S.H.; Chung, S.J.; Yi, Y.; Jun, S.; Kim, Y.K.; Jung, H.S. Comparative Analysis of Symmetry Parameters in the E2 Inner Core of the Pyruvate Dehydrogenase Complex. *Int. J. Mol. Sci.* **2024**, *25*, 13731. https://doi.org/10.3390/ijms252413731.
7. Hinostroza, F.; Albornoz-Muñoz, S.; Vergara, S.; Urra, G.; Araya-Durán, I.; Fissore, R.A.; González-Nilo, F.D.; Bustos, D.; Carvacho, I. Structural Implications of H233L and H398P Mutations in Phospholipase Cζ: A Full-Atom Molecular Dynamics Study on Infertility-Associated Dysfunctions. *Int. J. Mol. Sci.* **2025**, *26*, 4706. https://doi.org/10.3390/ijms26104706.
8. Wang, S.-N.; Liu, X.-X.; Lei, L.-J.; Wang, Q.; Shao, Z.-Q.; Liu, Y. Evolutionary Origins and Functional Diversification of 2′-O-Methyltransferases: Insights from Phylogenetic and Structural Analysis. *Int. J. Mol. Sci.* **2025**, *26*, 5260. https://doi.org/10.3390/ijms26115260.

References

1. Grandori, R. Protein Structure and Dynamics in the Era of Integrative Structural Biology. *Front. Biophys.* **2023**, *1*, 1219843. [CrossRef]
2. Lyon, A.S.; Peeples, W.B.; Rosen, M.K. A Framework for Understanding the Functions of Biomolecular Condensates across Scales. *Nat. Rev. Mol. Cell Biol.* **2021**, *22*, 215–235. [CrossRef] [PubMed]
3. Gellman, S.H. Introduction: Molecular Recognition. *Chem. Rev.* **1997**, *97*, 1231–1232. [CrossRef] [PubMed]
4. Carugo, O.; Djinović-Carugo, K. Structural Biology: A Golden Era. *PLoS Biol.* **2023**, *21*, e3002187. [CrossRef] [PubMed]
5. Shi, Y. A Glimpse of Structural Biology through X-Ray Crystallography. *Cell* **2014**, *159*, 995–1014. [CrossRef] [PubMed]
6. Lu, X.; Myint, W.; Muli, C.S.; Larion, M.; Matsuo, H.; Walters, K.J. NMR in the Age of Modern Biomedical Research and Drug Discovery. *J. Mol. Biol.* **2025**, *437*, 169302. [CrossRef] [PubMed]
7. Mu, X.; Gillman, C.; Nguyen, C.; Gonen, T. An Overview of Microcrystal Electron Diffraction (MicroED). *Annu. Rev. Biochem.* **2021**, *90*, 431–450. [CrossRef] [PubMed]
8. Nussinov, R.; Liu, Y.; Zhang, W.; Jang, H. Protein Conformational Ensembles in Function: Roles and Mechanisms. *RSC Chem. Biol.* **2023**, *4*, 850–864. [CrossRef] [PubMed]
9. Ellaway, J.I.J.; Anyango, S.; Nair, S.; Zaki, H.A.; Nadzirin, N.; Powell, H.R.; Gutmanas, A.; Varadi, M.; Velankar, S. Identifying Protein Conformational States in the Protein Data Bank: Toward Unlocking the Potential of Integrative Dynamics Studies. *Struct. Dyn.* **2024**, *11*, 034701. [CrossRef]
10. Nam, K.H. Guide to Serial Synchrotron Crystallography. *Curr. Res. Struct. Biol.* **2024**, *7*, 100131. [CrossRef]
11. Schmidt, M. Biological Function Investigated by Time-Resolved Structure Determination. *Struct. Dyn.* **2023**, *10*, 010901. [CrossRef]
12. Mäeots, M.-E.; Enchev, R.I. Structural Dynamics: Review of Time-Resolved Cryo-EM. *Acta Crystallogr. D Struct. Biol.* **2022**, *78*, 927–935. [CrossRef]
13. Caramello, N.; Royant, A. From Femtoseconds to Minutes: Time-Resolved Macromolecular Crystallography at XFELs and Synchrotrons. *Acta Crystallogr. D Struct. Biol.* **2024**, *80*, 60–79. [CrossRef]
14. Park, J.; Nam, K.H. Recent Chemical Mixing Devices for Time-Resolved Serial Femtosecond Crystallography. *TrAC Trends Anal. Chem.* **2024**, *172*, 117554. [CrossRef]
15. Hollingsworth, S.A.; Dror, R.O. Molecular Dynamics Simulation for All. *Neuron* **2018**, *99*, 1129–1143. [CrossRef] [PubMed]
16. Van Montfort, R.L.M.; Workman, P. Structure-Based Drug Design: Aiming for a Perfect Fit. *Essays Biochem.* **2017**, *61*, 431–437. [CrossRef] [PubMed]
17. Gonlepa, M.K.; Osotuyi, T.B.; Ofuonye, C.G.; Durojaye, O.A. Protein Engineering as a Driver of Innovation in Therapeutics Biotechnology and the Global Bioeconomy. *Discov. Chem.* **2025**, *2*, 226. [CrossRef]

https://doi.org/10.3390/ijms27020961

International Journal of
Molecular Sciences

MDPI

Article

Peculiarities of Scattering of Ultrashort Laser Pulses on DNA and RNA Trinucleotides

Dmitry Makarov * and Anastasia Kharlamova

Department of Fundamental and Applied Physics, Northern (Arctic) Federal University, Nab. Severnoi Dviny 17, 163002 Arkhangelsk, Russia
* Correspondence: makarovd0608@yandex.ru

Abstract: Currently, X-ray diffraction analysis (XRD) with high spatial and time resolution (TR-XRD) is based on the known theory of X-ray scattering, where the main parameter of USP—its duration—is not taken into account. In the present work, it is shown that, for scattering of attosecond USPs on DNA and RNA trinucleotides, the pulse length is the most important scattering parameter. The diffraction pattern changes considerably in comparison with the previously known scattering theory. The obtained results are extremely important in TR-XRD when using attosecond pulses to study trinucleotides of DNA and RNA, because with the previously known scattering theory, which does not take into account the duration of USP, one cannot correctly interpret, and therefore "decode", DNA and RNA structures.

Keywords: X-ray scattering; DNA; RNA; X-ray diffraction; ultrashort pulse; trinucleotides; CCG; CGG; CAG; CUG

1. Introduction

It is well known that X-ray diffraction analysis (XRD) is the primary method of determining the structure of matter [1–4]. The structures of most crystals and complex biomolecules, such as deoxyribonucleic acid (DNA) and ribonucleic acid (RNA), have been determined using this method. The theoretical basis for diffraction analysis dates back to the work of Max von Laue, who won the Nobel Prize in Physics in 1914 for the discovery of how X-rays can diffract on crystals [2]. The diffraction pattern obtained by scattering X-rays from various substances is the "imprint" of this material, whose deciphering enables the determination of crystal structures. The importance of this area can be estimated by the number of Nobel Prizes awarded for work related to X-ray crystallography, of which there are now more than 20. The key quantity in X-ray scattering is the scattering spectrum (determining the diffraction pattern) which is related to the Fourier transform of the electron density distribution $\rho(\mathbf{r})$ in matter as follows [5,6]

$$\frac{d\varepsilon}{d\Omega} = \frac{d\varepsilon_e}{d\Omega}\left|\rho(\mathbf{r})e^{i\mathbf{pr}}d^3r\right|^2, \quad (1)$$

where $\frac{d\varepsilon_e}{d\Omega}$ is the scattering spectrum of a free electron (Thomson scattering), $\mathbf{p}$ is the momentum transferred to the electron during light scattering (otherwise, $\mathbf{p} = \frac{2\pi}{\lambda}(\mathbf{n} - \mathbf{n}_0)$ is the scattering vector, where $\mathbf{n}$ is the direction of scattered radiation, $\mathbf{n}_0$ is the direction of the primary radiation), Ω is the solid angle at which scattering occurs. Using known methods [7] it is possible to determine $\rho(\mathbf{r})$ from the scattering pattern of X-rays. The surprising result is that Equation (1) can be derived from the classical description of electromagnetic radiation scattered by stationary electron density $\rho(\mathbf{r})$, which gives a result identical to that obtained with the quantum electrodynamic description of scattering [5]. The simplicity of the Equation (1) has given it wide distribution in XRD using not only conventional X-ray sources, but also sources generating ultrashort X-ray pulses (USPs). Further developments in physics have led to the creation and use of USPs to investigate the structure of matter. This is primarily related to the study of the structure of matter with not only high spatial but also temporal resolution. To study the structure of matter with high temporal resolution, the duration of the X-ray pulse must be shorter than the time scale on which the motion of the electron wave packet unfolds. In this case (e.g., [2,6,8]), instead of Equation (1)

$$\frac{d\varepsilon}{d\Omega} = \frac{d\varepsilon_e}{d\Omega}\left|\rho(\mathbf{r},t)e^{i\mathbf{p}\mathbf{r}}d^3r\right|^2. \quad (2)$$

Equation (2) differs from Equation (1) only by replacing $\rho(\mathbf{r}) \to \rho(\mathbf{r},t)$. This substitution provides access to the instantaneous electron density $\rho(\mathbf{r},t)$ as a function of time. This method is well known as time-resolved X-ray diffraction (TR-XRD). In other words the TR-XRD technique uses the same Equation (1), but the diffraction pattern is "read off" the system being studied over time t and conclusions are drawn about the dynamics of the system from a large set of such patterns.

Although one can see that the Equation (2) does not contain values characterizing USPs, this question deserves a separate consideration. Indeed, particular attention is now being paid to the physics of USPs, namely, the methods for generating such pulses and their interaction with matter [9–11]. Using such USPs, one can investigate the structure of complex substances with an advanced structure with high temporal and spatial resolution. Moreover, it is now technically possible to carry out such studies, e.g., using XFELs [12] free-electron lasers. Such lasers can generate USPs not only of femtosecond but also of attosecond duration [13,14]. Despite this, old theoretical approaches for "deciphering" polyatomic system structures based on the use of Equations (1) and (2), which do not take into account the features of interaction of such USPs with complex polyatomic structures are currently used [6,8,15–17].

Recently, it has been shown in [18–20] that there are important cases where the specific scattering of USPs must be accounted for in diffraction analysis; consequently, the use of Equations (1) and (2) results in significant errors. Basically, such errors appear in the scattering of attosecond USPs on certain kinds of structures. Such structures have been shown to include DNA and RNA nucleotides, etc. Thus, it is important to study structures where previously known diffraction analysis theory would give wrong results. By using such erroneous results when "deciphering" structures, it is possible to incorrectly identify the structures of the object under study. An interesting system to study scattering is DNA and RNA molecules. Let us look at some of the trinucleotides in DNA and RNA: CCG, CGG, CAG, CUG. Trinucleotides or triplets are repetitive regions in a molecule. The arrangement of triplets indicates the presence or absence of DNA and RNA mutations [21]. The mechanism of repeat formation is not fully known, but research in this area is actively pursued [22–25].

In this paper, it is shown that the USP duration for scattering on DNA and RNA trinucleotides: CCG, CGG, CAG, CUG, is one of the most important characteristics when using attosecond pulses. If the pulses are much longer than attosecond pulses, the duration parameter does not play a significant role in the scattering pattern and one can use Equations (1) or (2). The results show that the use of the Equation (1), on the basis of which complex molecules and biomolecules are "decoded", should be applied with great caution in the case of attosecond USPs.

Next, we will use the atomic system of units: $\hbar = 1$; $|e| = 1$; $m_e = 1$, where $\hbar$ is the Dirac constant, e is the electron charge, m_e is the electron mass.

2. Results and Discussion

As a basis, consider the DNA and RNA trinucleotides: CCG, CGG, CAG, CUG, on which the USP falls in the $\mathbf{n}_0$ direction, see Figures 1–4. Of course, the trinucleotides data presented in Figures 1–4 are modeled by us using their famous model. In reality, these trinucleotides can have an altered spatial structure, for example, compressed or stretched, including those without some constituent elements, i.e., atoms. Moreover, we can consider the variant when these structures are destroyed, i.e., when the atoms that make up these trinucleotides move according to some law $\delta\mathbf{R}(t) = \mathbf{v}t$, where v is the decay rate, and t is the decay time. We will not consider these cases, but will limit ourselves to the trinucleotides presented in Figures 1–4. This is entirely justified, since if the duration of the USP will affect the scattering spectra of the trinucleotides shown in Figures 1–4, then the length of the USP will play a big role in other cases related to these structures.

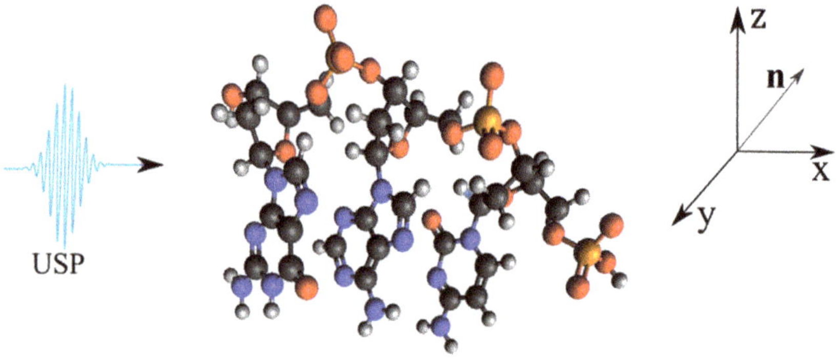

Figure 1. The CAG trinucleotide on which the USP falls is depicted, along with the selected coordinate system. The scattering spectra were calculated in the spatial orientation of the CAG with respect to the USP in the selected coordinate system shown in this figure. The red balls are oxygen atoms (O), the yellow ones are phosphorus (P), the blue ones are nitrogen (N), the light grey ones are hydrogen (H) and the dark grey ones are carbon (C).

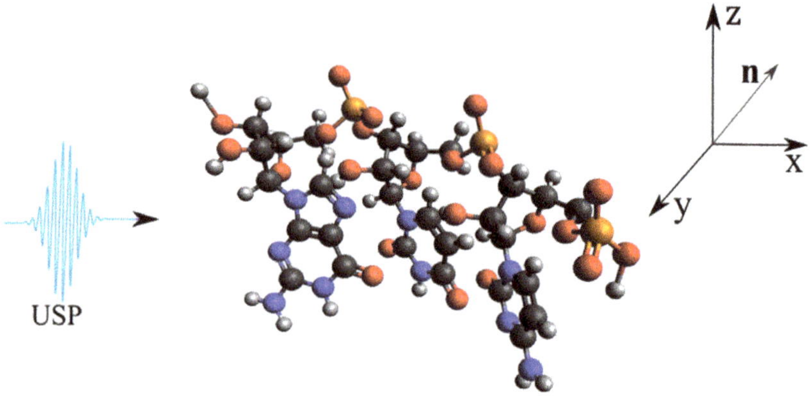

Figure 2. The same as in Figure 1, but only for trinucleide CUG.

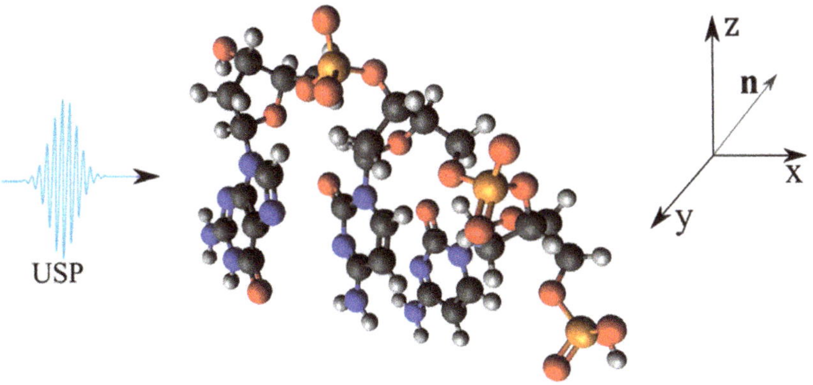

Figure 3. The same as in Figure 1, but only for trinucleide CCG.

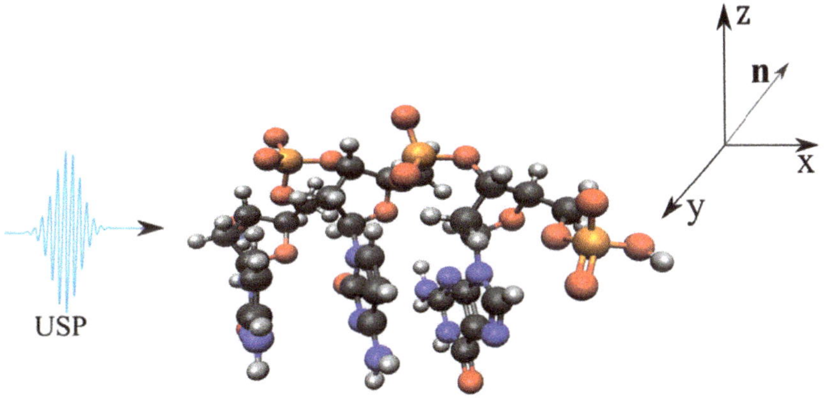

Figure 4. The same as in Figure 1, but only for trinucleide CGG.

We will use the sudden perturbation approximation to find the scattering spectra of USPs. It is assumed that the duration of the ultrashort pulse τ is many times shorter than the characteristic atomic time $\tau_a \sim 1$, i.e., $\tau \ll \tau_a$. As has been shown in [26], the condition of applicability of the sudden interaction approximation $\tau \ll \tau_a$ need not be fulfilled in the case of X-ray USPs. It has also been shown [26,27] that for X-ray USPs it is sufficient to assume that $\omega_0\tau_a \gg 1$, where ω_0 is the carrier frequency of incident USPs, to extend the sudden perturbation theory to the case of X-ray USPs. In other words, it is sufficient to consider USPs as multi-cycle. We will consider the USP as spatially inhomogeneous, i.e., the USP electromagnetic field strength should be chosen as $\mathbf{E}(\mathbf{r},t) = \mathbf{E}_0 h(t - \mathbf{n}_0\mathbf{r}/c)$, where $\mathbf{E}_0$ is the field amplitude, $h(t - \mathbf{n}_0\mathbf{r}/c)$ is the arbitrary function defining the USP shape, c is the light speed (in a. u. $c \approx 137$). Currently, the USPs intensities at the most powerful XFELs where matter structure studies are carried out do not exceed 10^{23} W/cm^2. In this case (more specifically, at $\mathrm{I} \ll 10^{25}$ W/cm^2), as shown in [18,19,27], the magnetic component of the USP can be neglected. In this case, the scattering spectra (scattered energy USPs per unit solid angle in the unit frequency interval) can be represented as [18,19,27,28]

$$\frac{d^2\varepsilon}{d\Omega_{\mathbf{k}}d\omega} = \frac{[\mathbf{E}_0\mathbf{n}]^2}{(2\pi)^2}\frac{|\tilde{h}(\omega)|^2}{c^3}\left[\sum_{i=1}^{s} N_{e,i}N_{A,i}(1-|F_i(\mathbf{p})|^2) + \sum_{i,j=1}^{s}\delta_{i,j}N_{e,i}N_{e,j}F_i(\mathbf{p})F_j^*(\mathbf{p})\right], \quad (3)$$

where $N_{A,i}$ is the number of atoms i variety; $N_{e,i}$ is the number of electrons in the atom i variety; summing over all atoms (i,j) up to s; $\tilde{h}(\omega) = \int_{-\infty}^{+\infty} h(\eta)e^{i\omega\eta}d\eta$, and $\mathbf{p} = \frac{\omega}{c}(\mathbf{n} - \mathbf{n}_0)$ has the meaning of recoil momentum when a USP is scattered on a bound electron; $F_i(\mathbf{p}) = \frac{1}{N_{e,i}}\int \rho_{e,i}(\mathbf{r})e^{-i\mathbf{pr}}d^3\mathbf{r}$ is the form factor of the i atom of the variety with electron density $\rho_{e,i}(\mathbf{r})$. In Equation (3), factor $\delta_{i,j} = \sum_{Ai,A'j} e^{-i\mathbf{p}(\mathbf{R}_{Ai}-\mathbf{R}_{A'j})}$ is a crucial quantity in USPs scattering, since it depends only on the position of atoms in space, i.e., on the coordinates of the atoms i of the variety (with number Ai), whose position in space is determined by the radius vector $\mathbf{R}_{Ai}$. In this expression, the coordinates of each of the atoms can be shifted by $\delta\mathbf{R}_{Ai}$, thereby simulating various cases from shifts and contractions to decays and synthesis of trinucleotides during TR-XRD, where $\delta\mathbf{R}_{Ai} = \delta\mathbf{R}_{Ai}(t)$. In this case, the factor $\delta_{i,j}(t) = \sum_{Ai,A'j} e^{-i\mathbf{p}(\mathbf{R}_{Ai}+\delta\mathbf{R}_{Ai}(t)-\mathbf{R}_{A'j}-\delta\mathbf{R}_{A'j}(t))}$. It is known that the case of main interest for XRD is at $\tau\omega_0 \gg 1$, i.e., USP, and is multicycle. If we make the assumption that $\tau \to \infty$, then we obtain the previously well known XRD theory where the radiation source is continuous. For the calculation, we will use the so-called Gaussian USP, which is the USP where the shape has a Gaussian envelope, i.e., $h(t) = e^{-i(\omega_0 t - \mathbf{k}_0\mathbf{r})}e^{-\alpha^2(t-\mathbf{n}_0\mathbf{r}/c)^2}$, where $\alpha = 1/\tau$, $\mathbf{k}_0 = \mathbf{n}_0\omega_0/c$. It should be added that such a USP is one of the best known for describing the shape of USPs and is often used in theoretical analysis. For example, in [29], based on exact solutions of the electromagnetic field equations, it was shown that at $\omega_0/\alpha \gg 1$, the solution takes the form of a simple Gaussian USP. In this case it is not difficult to show that $\tilde{h}(\omega) = \frac{\sqrt{\pi}}{\alpha}e^{-(\omega-\omega_0)^2/4\alpha^2}$. The condition $\omega_0/\alpha \gg 1$ is exactly the same as the multi-cycle USP mainly used in XRD. In this case it is not difficult to obtain the full scattering spectrum using Equation (3) in the form

$$\frac{d\varepsilon}{d\Omega_{\mathbf{k}}} = \frac{[\mathbf{E}_0\mathbf{n}]^2}{4c^3\alpha\sqrt{2\pi}}\left[\sum_{i=1}^{s} N_{e,i}N_{A,i}(1 - |F_i(\mathbf{p}_0)|^2) + \sum_{i,j=1}^{s} \gamma_{i,j}(\mathbf{p}_0, \mathbf{p}_\tau)N_{e,i}N_{e,j}F_i(\mathbf{p}_0)F_j^*(\mathbf{p}_0)\right],$$

$$\gamma_{i,j}(\mathbf{p}_0, \mathbf{p}_\tau) = \sum_{Ai,A'j} e^{-i\mathbf{p}_0(\mathbf{R}_{Ai}-\mathbf{R}_{A'j}))}e^{-\frac{1}{2}(\mathbf{p}_\tau(\mathbf{R}_{Ai}-\mathbf{R}_{A'j}))^2}. \quad (4)$$

where $F_i(\mathbf{p}_0), \delta_{i,j}(\mathbf{p}_0)$ at $\mathbf{p}_0 = \frac{\omega_0}{c}(\mathbf{n} - \mathbf{n}_0)$ and $\mathbf{p}_\tau = \frac{1}{c\tau}(\mathbf{n} - \mathbf{n}_0)$. The electron density of the atoms $\rho_{e,i}$ variety i will be chosen in the independent atom model [30]. In this case, we get $\rho_{e,i}(\mathbf{r}) = \frac{N_{e,i}}{4\pi r}\sum_{k=1}^{3} A_{k,i}\alpha_{k,i}^2 e^{-\alpha_{k,i}r}$, where $A_{k,i}, \alpha_{k,i}$ are constant coefficients defined in [30].

Equation (4) contains characteristics which are responsible for the duration of the τ USP, and hence the specificity of the USP scattering on the system under study is taken into account. In doing so, one can see that if in Equation (4) we increase $\tau \to \infty$, we get the well known Equation (1) or (2) in case of dependence of atomic coordinates on time, see, e.g., [19]. This is not difficult to show, if $\tau \to \infty$ with $\mathbf{p}_\tau = \frac{1}{c\tau}(\mathbf{n} - \mathbf{n}_0) \to 0$; thus, Equation (4) becomes proportional to τ (Fermi's golden rule). And Equation (1) in this case, exactly corresponds to the golden Fermi rule. It can be shown qualitatively that the pulse duration τ becomes an important scattering parameter only for attosecond and shorter pulses. For this purpose one can see that in the parameter $\gamma_{i,j}(\mathbf{p}_0, \mathbf{p}_\tau)$ in Equation (4) there is an important factor $e^{-\frac{1}{2}(\mathbf{p}_\tau(\mathbf{R}_{Ai}-\mathbf{R}_{A'j}))^2}$ which is close to unity for small $\mathbf{p}_\tau(\mathbf{R}_{Ai} - \mathbf{R}_{A'j}) \ll 1$. A small value of $\mathbf{p}_\tau(\mathbf{R}_{Ai} - \mathbf{R}_{A'j}) \ll 1$ will almost always be the case if the pulses are many times longer than attosecond pulses, since in this case $\mathbf{p}_\tau \lesssim 10^{-4}$. If we choose shorter attosecond pulses, then $\mathbf{p}_\tau \gtrsim 10^{-2}$, which leads to considerable values of $\mathbf{p}_\tau(\mathbf{R}_{Ai} - \mathbf{R}_{A'j})$ for multi-atomic systems. In this case, the duration of the USP cannot be neglected. Of course, this is a qualitative analysis and one can only say more accurately about the effect of pulse duration after calculations using Equation (4). Although one can quite confidently state that the duration of USPs must be taken into account only for attosecond and shorter pulses. It is also possible to explain qualitatively in terms of physics why the duration of USPs can significantly affect the scattering spectra and why Equation (1) is not always correct. When some fraction of atoms in the matter fall into a spatial region of USP size ($\sim c\tau$), the emission is coherent from this region. Accordingly, if the duration of the USP is large, i.e., $\tau \gg 1$, then the region where emission occurs coherently is also large. In Equation (1), such a region where radiation occurs coherently is infinitely large. This will indeed be true in the case of USP when the region ($\sim c\tau$) captures the whole system (molecule, nanosystem, etc.) under study. If, on the contrary, when the $\sim c\tau$ region captures a smaller region than the size of the system being studied, there will be no coherent emission from that whole system. That is why the study of USP scattering by bimolecules, especially DNA and RNA, will be relevant, since such molecules are quite large and the size of the attosecond pulse $\sim c\tau$ becomes smaller than the size of the molecule itself. This is a very important refinement because when using attosecond pulses in XRD one must take into account the duration of the USP, which is not taken into account in Equation (1). This in turn leads to incorrect determination of the matter structure using Equation (1).

Let us perform USP scattering calculations on the systems shown in Figures 1–4 and show that for such systems the use of Equation (1) introduces substantial errors. Calculations of scattering spectra using Equation (4) are shown in Figures 5–8. In the calculations, the USP was assumed to fall as shown in Figures 1–4 with photon energy $\hbar\omega_0 = 7.46$ keV, pulse duration $\tau = 1$ (as). It should be added that the choice of the duration of the USP $\tau = 1$ (as) purely conditional, for a better reference in the timeline. If we choose the duration of the tens $\tau \sim 10$ (as) the results are close, with the only difference being that they are more similar to the results of Equation (1). If we take into account that USPs with a duration of tens of attoseconds [13,14] are already realized, then our theory is feasible in practice.

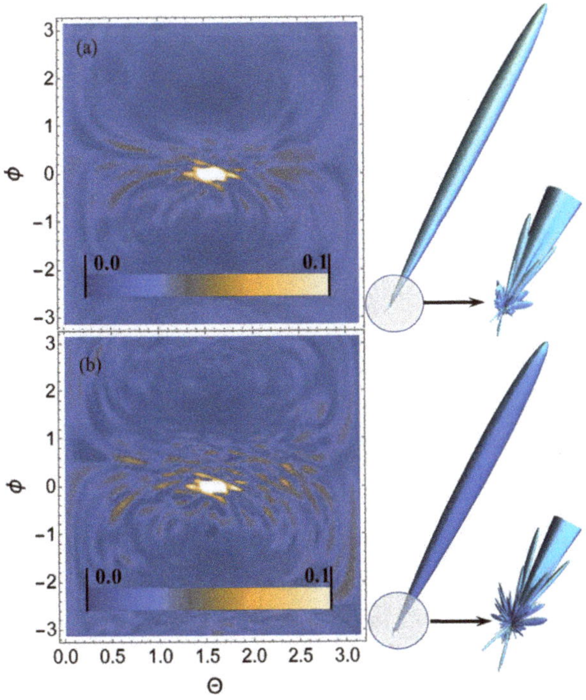

Figure 5. USP scattering spectra on the CAG trinucleotide: (**a**) calculated using Equation (4), (**b**) calculated by Equation (1). The spectra are presented as 2D (**left**) and 3D (**right**) plots. The 2D plots are presented in dimensionless units and normalised to the maximum value of the spectrum. The angles ϕ, θ are spherical angles in the coordinate system shown in Figure 1.

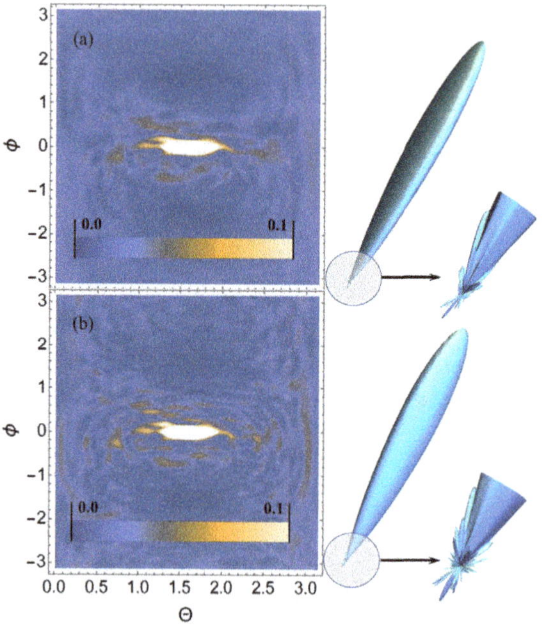

Figure 6. USP scattering spectra on the CUG trinucleotide: (**a**) calculated using Equation (4), (**b**) calculated by Equation (1). The spectra are presented as 2D (**left**) and 3D (**right**) plots. The 2D plots are presented in dimensionless units and normalised to the maximum value of the spectrum. The angles ϕ, θ are spherical angles in the coordinate system shown in Figure 2.

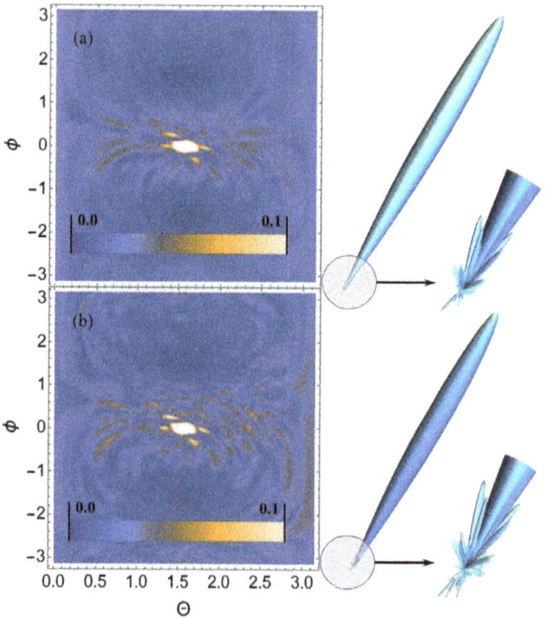

Figure 7. USP scattering spectra on the CCG trinucleotide: (**a**) calculated using Equation (4), (**b**) calculated by Equation (1). The spectra are presented as 2D (**left**) and 3D (**right**) plots. The 2D plots are presented in dimensionless units and normalised to the maximum value of the spectrum. The angles ϕ, θ are spherical angles in the coordinate system shown in Figure 3.

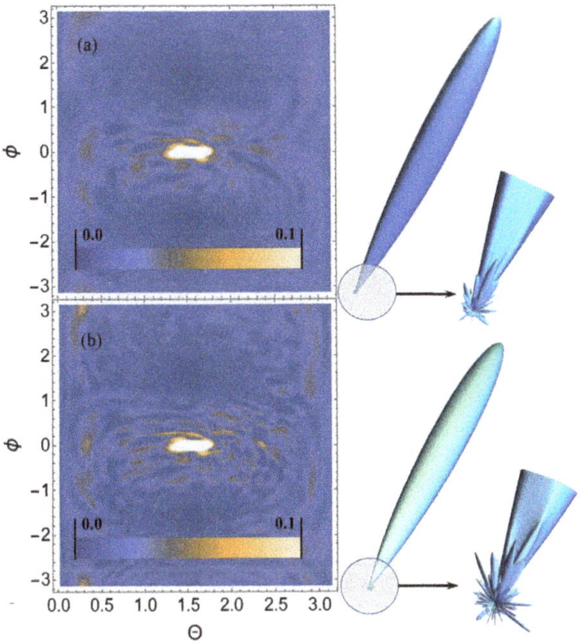

Figure 8. USP scattering spectra on the CGG trinucleotide: (**a**) calculated using Equation (4), (**b**) calculated by Equation (1). The spectra are presented as 2D (**left**) and 3D (**right**) plots. The 2D plots are presented in dimensionless units and normalised to the maximum value of the spectrum. The angles ϕ, θ are spherical angles in the coordinate system shown in Figure 4.

From the Figures 5–8 we can see that the scattering spectra in the case of using Equations (1) and (4) are significantly different. The central peaks, no matter how the calculations were carried out, are always close to each other and this is well explained, since this peak is directed in the direction of the incident USP $\mathbf{n}_0$. Indeed, it is quite obvious that most of the USP will scatter in the direction $\mathbf{n}_0$. Other diffraction peaks already differ essentially from each other, and when scattering to more angles they differ even qualitatively, i.e., the scattering is different. At this point, it is possible to notice that the same regularity at scattering, irrespective of investigated trinucleotides, is the reduction of peaks in a diffraction picture at the big angles of scattering in case of using Equation (4). This reduction of peaks in the diffraction pattern can be explained by the influence of the pulse duration on the scattering spectra.

As a result, comparing the results of calculations and analysis of Equation (4), we can draw an important conclusion. In the case of attosecond USPs scattering on DNA and RNA trinucleotides: CCG, CGG, CAG, CUG, the duration of the USPs is an important parameter in scattering, and to use the previously known Equation (1) in XRD or Equation (2) in TR-XRD is not correct. It should be added that this conclusion would have even more significance if we consider that the use of Equation (1) in XRD or Equation (2) in TR-XRD has been standard for many years. Our approach, presented here, for a sufficiently long USP and a multi-atom system, passes to the previously known theory, i.e., Equation (4) becomes equal to Equation (1) in XRD or Equation (2) in TR-XRD. Thus Equation (4) is more general in XRD theory (or TR-XRD), where the previous theory is a special case of the theory presented here. Obviously, in the case of using attosecond pulses, as explained above, to determine structures of more complex molecules than those considered here, the conclusions may be the same. Of course, it is technically quite difficult to study trinucleotides separately, since they are part of DNA and RNA. One way or another, scattering on the basic elements of DNA and RNA (which are trinucleotides periodically repeating in space) gives its own spectrum, and the periodicity from trinucleotides in space in DNA and RNA molecules gives a different spectrum [18,19]. As a result, if the scattering spectra from trinucleotides, using our theory, will differ from the scattering spectra calculated from Equations (1) or (2), then the scattering spectra for the entire DNA or RNA molecule will also differ. Thus, the calculation data obtained and the analysis of Equations (3) and (4) indicate the importance of the conclusions presented here. The main conclusion is the incorrectness of using the previously known XRD or TR-XRD theory in the case of attosecond pulses, where for such pulses it is necessary to use Equations (3) and (4), which take into account the pulse duration parameter. We should also add that here we studied scattering on spatially oriented trinucleotides, which is not the case in the experiment using the pump-probe technique. It is well known that this method has a significant advantage over the standard XRD method [8]. The sample in this method is real, i.e., no additional preparatory procedures are carried out on it, after which the sample may change significantly. It is this method that makes it possible to monitor changes in the structure of complex polyatomic systems in real time. In this technique, the molecules under study are fed chaotically (i.e., random) oriented with respect to the USP. Despite this, it is obvious that if the results in our theory differ from Equations (1) or (2) with a given orientation, then they will also differ with chaotic orientation. This can be clearly seen from Equation (4), where the duration parameter does not disappear when averaged over the orientation of the molecule. Of course, it is quite interesting to give the results of calculations under chaotic orientation of trinucleotides, which we also did, and the results calculated by Equations (1) and (4) are indeed also significantly different as in Figures 1–4.

3. Materials and Methods

As the methods of our research, we have chosen the theory here developed, see Equations (3) and (4) based on [18,19,26,28]. This method is based on the sudden perturbation approximation [26] and the theory of scattering of ultrashort electromagnetic field pulses [28]. The basic technique for calculating scattering spectra on multiatomic complex systems is presented in [18,19]. Trinucleotides DNA and RNA were chosen as the object of the study: CCG, CGG, CAG, CUG.

4. Conclusions

Thus, we can conclude that the theory presented here (see Equation (4)) is needed in TR-XRD to study processes in trinucleotides occurring on an atomic and larger time scale ($\tau_a \sim 100$ as), in the case of using attosecond pulses with a duration of $\tau \ll \tau_a$ where Equation (2) gives errors (previously known theory). Also, to study the structure of trinucleotides in XRD, you need to know that when using attosecond pulses, Equation (1) gives errors, and in this case, to determine the structure, it is better and easier to use longer pulses, for example, femtosecond and Equation (1). Although this work is theoretical, it has a very clear practical orientation—it is the deciphering of complex polyatomic structures in both stationary and dynamic states.

Author Contributions: Conceptualization, D.M.; methodology, D.M.; software, D.M. and A.K.; validation, D.M. and A.K.; formal analysis, D.M. and A.K.; writing—original draft preparation, D.M. and A.K.; writing—review and editing, D.M.; project administration, D.M. All authors have read and agreed to the published version of the manuscript.

Funding: The research was supported by Grant of the President of the Russian Federation No. MD-4260.2021.1.2; state assignment of the Russian Federation No. 0793-2020-0005 and No. FSRU-2021-0008.

Institutional Review Board Statement: Not applicable.

Informed Consent Statement: Not applicable.

Data Availability Statement: Request to corresponding author of this article.

Conflicts of Interest: The authors declare no conflict of interest.

References

1. Suryanarayana, C.; Grant, N.M. *X-ray Diffraction: A Practical Approach*; Plenum Press: New York, NY, USA; London, UK, 1998.
2. Jones, N. Crystallography: Atomic secrets. *Nature* **2018**, *505*, 602–603. [CrossRef] [PubMed]
3. Pietsch, U.; Holy, V.; Baumbach, T. *High-Resolution X-ray Scattering*; Springer Science & Business Media: New York, NY, USA, 2004.
4. Benediktovich, A.; Feranchuk, I.; Ulyanenkov, A. *Theoretical Concepts of X-ray Nanoscale Analysis*; Springer: Berlin/Heidelberg, Germany, 2014.
5. James, R. *The Optical Principles of the Diffraction of X-rays (Ox Bow)*; Ox Bow Press: Woodbridge, UK, 1982.
6. Dixit, G.; Vendrell, O.; Santra, R. Imaging electronic quantum motion with light. *Proc. Natl. Acad. Sci. USA* **2012**, *109*, 11636–11640. [CrossRef] [PubMed]
7. Hauptman, H.A. The phase problem of x-ray crystallography. *Rep. Prog. Phys.* **1991**, *54*, 1427–1454. [CrossRef]
8. Eseev, M.K.; Matveev, V.I.; Makarov, D.N. Diagnostics of Nanosystems with the Use of Ultrashort X-ray Pulses: Theory and Experiment (Brief Review). *JETP Lett.* **2021**, *114*, 387–405. [CrossRef]
9. Krausz, F.; Ivanov, M. Attosecond physics. *Rev. Mod. Phys.* **2009**, *81*, 163. [CrossRef]
10. Kraus, P.; Zürch, M.; Cushing, S.K.; Neumark, D.M.; Leone, S.R. The ultrafast X-ray spectroscopic revolution in chemical dynamics. *Nat. Rev. Chem.* **2018**, *2*, 82–94. [CrossRef]
11. Peng, P.; Marceau, C.; Villeneuve, D.M. Attosecond imaging of molecules using high harmonic spectroscopy. *Nat. Rev. Phys.* **2019**, *1*, 144–155. [CrossRef]
12. Schoenlein, R.; Elsaesser, T.; Holldack, K.; Huang, Z.; Kapteyn, H.; Murnane, M.; Woerner, M. Recent advances in ultrafast X-ray sources. *Philos. Trans. R. Soc. A* **2019**, *377*, 20180384. [CrossRef]
13. Duris, J.; Li, S.; Driver, T.; Champenois, E.G.; MacArthur, J.P.; Lutman, A.A.; Marinelli, A. Tunable isolated attosecond X-ray pulses with gigawatt peak power from a free-electron laser. *Nat. Photonics* **2020**, *14*, 30–36. [CrossRef]
14. Maroju, P.K.; Grazioli, C.; Di Fraia, M.; Moioli, M.; Ertel, D.; Ahmadi, H.; Sansone, G. Attosecond pulse shaping using a seeded free-electron laser. *Nature* **2020**, *578*, 386–391. [CrossRef]
15. Leone, S.R.; McCurdy, C.W.; Burgdörfer, J.; Cederbaum, L.S.; Chang, Z.; Dudovich, N.; Vrakking, M.J. What will it take to observe processes in "real time"? *Nat. Photonics* **2014**, *8*, 162–166. [CrossRef]
16. Astapenko, V.A.; Sakhno, E.V. Excitation of a quantum oscillator by short laser pulses. *Appl. Phys. B* **2020**, *126*, 23. [CrossRef]
17. Rosmej, F.; Astapenko, V.; Lisitsa, V.; Li, X.; Khramov, E. Scattering of ultrashort laser pulses on "ion-sphere" in dense plasmas. *Contrib. Plasma Phys.* **2019**, *59*, 189–196. [CrossRef]
18. Makarov, D.; Kharlamova, A. Scattering of X-ray Ultrashort Pulses by Complex Polyatomic Structures. *Int. J. Mol. Sci.* **2022**, *23*, 163. [CrossRef]
19. Makarov, D.N.; Makarova, K.A.; Kharlamova, A.A. Specificity of scattering of ultrashort laser pulses by molecules with polyatomic structure. *Sci. Rep.* **2022**, *12*, 4976. [CrossRef] [PubMed]
20. Eseev, M.; Makarova, K.; Makarov, D. Scattering of Ultrashort X-ray Pulses on Diamonds with NV Centers. *Crystals* **2022**, *12*, 1417. [CrossRef]
21. Cardoso, I.; Marques, V. Trinucleotide repeat diseases—Anticipation diseases. *J. Clin. Gen. Genom.* **2018**, *1*, 4–9.
22. Nakamori, M.; Panigrahi, G.B.; Lanni, S.; Gall-Duncan, T.; Hayakawa, H.; Tanaka, H.; Pearson, C.E. A slipped-CAG DNA-binding small molecule induces trinucleotide-repeat contractions in vivo. *Nat. Genet.* **2020**, *5252*, 146–159. [CrossRef]
23. Hu, T.; Morten, M.J.; Magennis, S.W. Conformational and migrational dynamics of slipped-strand DNA three-way junctions containing trinucleotide repeats. *Nat. Commun.* **2021**, *1212*, 204. [CrossRef] [PubMed]
24. Lee, J.M.; Correia, K.; Loupe, J.; Kim, K.H.; Barker, D.; Hong, E.P.; Myers, R.H. CAG Repeat Not Polyglutamine Length Determines Timing of Huntington's Disease Onset. *Cell* **2019**, *178*, 887–900. [CrossRef]
25. Tabrizi, S.J.; Flower, M.D.; Ross, C.A.; Wild, E.J. Huntington disease: New insights into molecular pathogenesis and therapeutic opportunities. *Nat. Rev. Neurol.* **2020**, *1616*, 529–546. [CrossRef] [PubMed]
26. Makarov, D.; Eseev, M.K.; Makarova, K.A. Analytical wave function of an atomic electron under the action of a powerful ultrashort electromagnetic field pulse. *Opt. Lett.* **2019**, *44*, 3042–3045. [CrossRef] [PubMed]
27. Eseev, M.; Goshev, A.; Makarova, K.; Makarov, D. X-ray diffraction analysis of matter taking into account the second harmonic in the scattering of powerful ultrashort pulses of an electromagnetic field. *Sci. Rep.* **2021**, *11*, 3571. [CrossRef] [PubMed]
28. Makarov, D. Quantum theory of scattering of ultrashort electromagnetic field pulses by polyatomic structures. *Opt. Express* **2019**, *27*, 31989–32008. [CrossRef]
29. Lin, Q.; Zheng, J.; Becker, W. Subcycle pulsed focused vector beams. *Phys. Rev. Lett.* **2006**, *97*, 253902. [CrossRef]
30. Salvat, F.; Martnez, J.; Mayol, R.; Parellada, J. Analytical Dirac-Hartree-Fock-Slater screening function for atoms (Z = 1–92). *Phys. Rev. A* **1987**, *36*, 467–474. [CrossRef]

International Journal of
Molecular Sciences

MDPI

Article

Structural Basis of the Inhibition of L-Methionine γ-Lyase from *Fusobacterium nucleatum*

Tingting Bu [1,2,†], Jing Lan [1,2,†], Inseong Jo [3], Jie Zhang [1,2], Xue Bai [1,2], Shanru He [1,2], Xiaoling Jin [1], Lulu Wang [1,2], Yu Jin [1,2], Xiaoyu Jin [1,2], Liying Zhang [1,2], Hailong Piao [4], Nam-Chul Ha [5], Chunshan Quan [1,2], Ki Hyun Nam [6,7,*] and Yongbin Xu [1,2,*]

1 Department of Bioengineering, College of Life Science, Dalian Minzu University, Dalian 116600, China
2 Key Laboratory of Biotechnology and Bioresources Utilization of Ministry of Education, College of Life Science, Dalian Minzu University, Dalian 116600, China
3 Infectious Diseases Therapeutic Research Center, Korea Research Institute of Chemical Technology, Daejeon 34114, Republic of Korea
4 CAS Key Laboratory of Separation Science for Analytical Chemistry, Dalian Institute of Chemical Physics, Chinese Academy of Sciences, Dalian 116600, China
5 Department of Agricultural Biotechnology, College of Agriculture and Life Sciences, Seoul National University, Gwanak-gu, Seoul 00826, Republic of Korea
6 Department of Life Science, Pohang University of Science and Technology, Pohang 35398, Republic of Korea
7 POSTECH Biotech Center, Pohang University of Science and Technology, Pohang 35398, Republic of Korea
* Correspondence: structures@postech.ac.kr (K.H.N.); yongbinxu@dlnu.edu.cn (Y.X.)
† These authors contributed equally to this work.

Abstract: *Fusobacterium nucleatum* is a lesion-associated obligate anaerobic pathogen of destructive periodontal disease; it is also implicated in the progression and severity of colorectal cancer. Four genes (*FN0625*, *FN1055*, *FN1220*, and *FN1419*) of *F. nucleatum* are involved in producing hydrogen sulfide (H_2S), which plays an essential role against oxidative stress. The molecular functions of Fn1419 are known, but their mechanisms remain unclear. We determined the crystal structure of Fn1419 at 2.5 Å, showing the unique conformation of the PLP-binding site when compared with L-methionine γ-lyase (MGL) proteins. Inhibitor screening for Fn1419 with L-cysteine showed that two natural compounds, gallic acid and dihydromyricetin, selectively inhibit the H_2S production of Fn1419. The chemicals of gallic acid, dihydromyricetin, and its analogs containing trihydroxybenzene, were potentially responsible for the enzyme-inhibiting activity on Fn1419. Molecular docking and mutational analyses suggested that Gly112, Pro159, Val337, and Arg373 are involved in gallic acid binding and positioned close to the substrate and pyridoxal-5′-phosphate-binding site. Gallic acid has little effect on the other H_2S-producing enzymes (Fn1220 and Fn1055). Overall, we proposed a molecular mechanism underlying the action of Fn1419 from *F. nucleatum* and found a new lead compound for inhibitor development.

Keywords: *Fusobacterium nucleatum*; hydrogen sulfide; L-methionine γ-lyase; L-cysteine desulfidase; Fn1419; gallic acid

1. Introduction

Anaerobic bacteria are the most common microflora of human beings and do not cause infections in the healthy host [1]. However, upon injury or trauma to the body, these bacteria cause infections that are typically suppurative. The rising antimicrobial resistance of anaerobic bacteria poses a significant threat to public health [2].

The formation of reactive oxygen species (ROS) was considered a common effector mechanism of bactericidal antibiotics [3]. Hydrogen sulfide (H_2S) plays an essential role in combatting ROS through its antioxidant effects, which can protect the body from oxidative stress-induced diseases [4,5]. In addition, H_2S biogenesis is a critical contributor to bacterial antibiotic tolerance and a target for multifunctional antibiotic synergists [6].

Understanding the modes of action of antibiotics and bacterial resistance mechanisms is especially important for developing alternative therapies.

The gram-negative anaerobic bacterium *Fusobacterium nucleatum* is associated with lesions in destructive periodontal disease [7]. It has also been related to the progression and severity of colorectal cancer [8]. Although it is linked to the pathogenesis of periodontal disease, it has attracted the most attention in colonic health, and the extent to which it is detrimental or beneficial to the host remains debatable. H_2S is produced in the gut from cysteine by epithelial cells and the intestinal microbiota [9]. Biological H_2S is produced primarily from cysteine and/or homocysteine by cystathionine β–synthase, cystathionine γ-lyase (CGL), cysteine aminotransferase, and 3-mercaptopyruvate sulfur transferase, which work either individually or in concert [10–13]. It should be noted that *F. nucleatum* produces high amounts of H_2S [14,15].

F. nucleatum harbors four genes (*FN0625, FN1055, FN1220, and FN1419*) encoding enzymes involved in the production of H_2S [16]. *FN0625* encodes aspartate aminotransferase, a pyridoxal 5′-phosphate (PLP)-dependent enzyme, which catalyzes the conversion of L-cysteine and H_2O to H_2S, pyruvate, and ammonium by an α, β-elimination reaction [16]. *FN1055* also encodes a PLP-dependent enzyme, a cysteine synthase, which catalyzes the conversion of L-cysteine and H_2O to H_2S and L-serine by a β-replacement reaction, an unusual cysteine (hydroxyl) lyase reaction. On the other hand, *FN1220* encodes a fold-type II PLP-dependent enzyme, a lanthionine synthase, which condenses two L-cysteine molecules to generate H_2S and the uncommon amino acid L-lanthionine through a β-replacement reaction [17]. Fn1419 has been characterized as an L-methionine γ-lyase (MGL^{Met}), which belongs to the γ-family of PLP-dependent enzymes and deaminates L-methionine to 2-oxobutanoate, methanethiol, and ammonia cation [16]. Fn1419 also uses L-cysteine (MGL^{Cys}) to produce H_2S with pyruvate and ammonia by α, β-elimination [16]. MGL enzymes are found in prokaryotic and eukaryotic pathogenic microorganisms but are absent from mammals [18]. Accordingly, Fn1419 could serve as a target for novel antibacterial drugs against *F. nucleatum*, i.e., for its inhibitors and/or suicidal substrates. However, the molecular mechanism underlying Fn1419 activity remains largely uncharacterized.

To better understand the molecular mechanism underlying the activity of Fn1419 to guide its inhibitor design, in the present study, we characterized the MGL^{Cys} activity of Fn1419 and determined its crystal structure at a resolution of 2.5 Å. An inhibition assay using 39 natural compounds revealed that gallic acid and dihydromyricetin inhibit the MGL^{Cys} activity of Fn1419. We also investigated the gallic acid-binding site using molecular docking and mutagenesis studies. These findings will improve our understanding of the molecular function of Fn1419 and provide insights into the design of its inhibitors.

2. Results

2.1. Characterization of Fn1419

Fn1419 was overexpressed in *Escherichia coli* and exhibited the tetrameric state on analytical size-exclusion chromatography (Figure 1a). After purification, the concentrated Fn1419 solution showed a yellowish color (Figure 1a). Since the MGL enzyme requires a PLP molecule as a cofactor [19,20], we considered that the yellow color could be the result of PLP binding to Fn1419. UV-Vis absorption of PLP solution and purified Fn1419 were analyzed to verify the binding of PLP to Fn1419. The PLP and Fn1419 solution showed absorption peaks at 420 nm and 425 nm, respectively. The wavelength of the absorption peak of Fn1419 was similar to the absorbance of PLP bound to other enzymes [21] (Figure 1b). Since PLP molecules were not added during protein expression and purification, Fn1419-bound PLP molecules were derived from *E. coli* during recombinant protein expression. Fn1419 can use L-cysteine to produce H_2S with pyruvate and ammonia by α, β-elimination [16] (Figure 1c). To verify the MGL^{Cys} activity of recombinant Fn1419 used in this experiment, an enzyme activity assay was performed using L-cysteine as a substrate. As a result, H_2S production was clearly observed by the reaction of Fn1419 (Figure 1d).

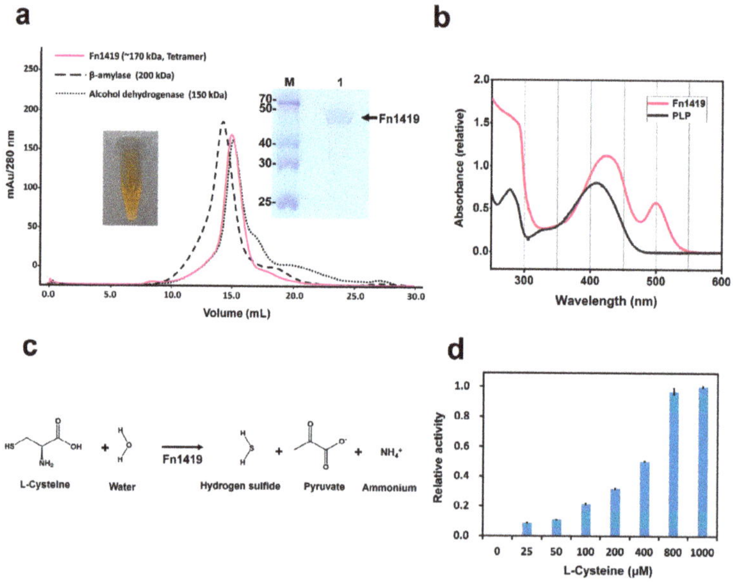

Figure 1. Characterization of the PLP-binding and MGL^{Cys} activity of Fn1419. (**a**) Profile of size-exclusion chromatography of Fn1419, showing the tetramer state in solution. (insert, left) The concentrated Fn1419 solution with yellow coloration. (insert, right) Purified Fn1419 on SDS-PAGE. (**b**) Spectroscopic analysis of Fn1419. The PLP and PLP-bound Fn1419 exhibited absorption peaks at 420 and 425 nm, respectively. (**c**) Reaction mechanism of *F. nucleatum* Fn1419 with L-cysteine. (**d**) The MGL^{Cys} activity of purified Fn1419 with different concentrations of L-cysteine. The error bars represent the standard deviation of three replicates.

2.2. Crystal Structure of Fn1419

To better understand the molecular function of Fn1419, its crystal structure was determined. The Fn1419 crystal belongs to the trigonal space group $P3_121$, with unit-cell parameters of a = b = 95.2 Å, and c = 302.3 Å. The final model was refined to 2.5 Å resolution, with R_{work} and R_{free} of 21.6% and 27.9%, respectively (Table 1). The electron density maps of Fn1419 were well-defined for all amino acid residues, except for the disordered region between Ser39 and Gly60.

The Fn1419 monomer consists of three spatially and functionally distinct subdomains: an N-terminal domain (NTD, residues Lys5–Asp38), a PLP-binding domain (PBD, residues Asn61–Leu245), and a C-terminal domain (CTD, residues Gly246–Ile395) (Figure 2a). The NTD consists of two α-helices (α1–α2) and is involved in dimer formation. The PBD consists of a central seven-stranded β-sheet (β1–β7), which is surrounded by seven α-helices (α3–α9), and one antiparallel β-strand (β7) that forms a β-hairpin structure with β6 (Figure 2a). The CTD consists of six α-helices (α10–α15) and four β-strands (β8–β11), forms two-antiparallel β-sheets, and is involved in catalysis. Further, the B-factor of the CTD (59.50 $Å^2$) is much higher than that of NTD (51.10 $Å^2$) and PBD (45.79 $Å^2$), indicating that the structures of NTD and PBD are more rigid than that of CTD. Of note, the B-factor of the loop between α13 and α14 of the CTD was much higher than that for any other protein segment (Figure S1). Therefore, we concluded that the CTD might be essential in recognizing and trapping the substrate.

Table 1. Crystallographic data and refinement statistics.

Data collection	**Fn1419**
Beamline	Beamline BL19U1 at SSRF
Resolution range (Å)	42.99–2.50 (2.589–2.5)
Space group	$P3_121$
a, b, c (Å)	95.21, 95.21, 302.25
α, β, γ (o)	90.00, 90.00, 120.00
R_{sym} (%)	6.2 (30.7)
Completeness (%)	99.55 (98.89)
Multiplicity	2.1 (2.1)
Average I/σ(I)	17.27 (3.15)
Refinement	
R_{work}/R_{free} (%)	21.60 (24.61)/27.88 (35.42)
Protein residues	395
Protein	52.73
Water	46.92
R.m.s.d. from ideal	
RMS (Bond)	0.009
RMS (Angles)	1.240
Ramachandran plot (%)	
Favored regions	94.73
Allowed regions	4.69
Disallowed regions	0.58
PDB code	7BQW

Values in parentheses are for the outermost shell.

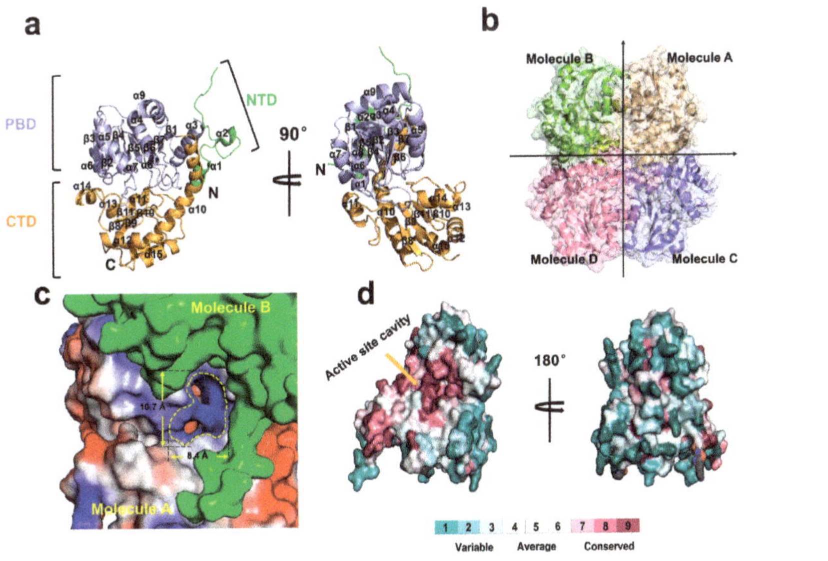

Figure 2. Analysis of the overall structure of Fn1419. (**a**) Monomer structure of Fn1419 consisting of the N-terminal (NTD), pyridoxal-5-phosphate–binding (PBD), and C-terminal (CTD) domains. (**b**) Tetrameric structure of Fn1419. (**c**) Electrostatistic surface of the dimer interface showing the active site. (**d**) Surface conservation of Fn1419 and MGL proteins. The active site showed high conservation, whereas the other regions were not conserved.

Fn1419 monomers form a tetrameric structure with an approximate dimension of 71 × 71 × 34 Å (Figure 2b), which is consistent with size-exclusion chromatography (Figure 1a). The dimer interface of Fn1419 molecules A and B is stabilized by 25 hydrogen bond interactions (Figure S2a and Table S1). While the total surface area of the Fn1419 monomer is 16,149 Å^2, the buried surface of monomers A and B is 1733 Å^2 and 1969 Å^2, respectively, of which 11.1% corresponds to the partner molecule. Further, eight hydrogen bond interactions are observed in the dimer interface between Fn1419 molecules A and D (Figure S2b and Table S1), with buried surface areas of 693 Å^2 and 650 Å^2 (for monomers A and D, respectively), of which 4.3% corresponds to the partner molecule. Finally, 38 hydrogen bonds and 19 salt bridge interactions are observed in the dimer interface between Fn1419 molecules A and C (Figure 2c and Table S1), with buried surface areas of 1516 Å^2 and 1510 Å^2 (for monomers A and C, respectively), of which 9.4% corresponds to the partner molecule. Further, catalytic dimeric interface analysis revealed that any two monomers contact each other via large, predominantly flat regions. At the upper dimeric interface, residues are mainly distributed in six loops (loop1, loop3, loop5, loop14, loop16, and loop21) and five α-helices (α1, α4, α9, α13, and α15). The Fn1419 dimer forms a positive-charged curtain-shaped active site pocket in the PBD of two monomers, approximately 10.7 Å long and 8.4 Å wide (Figure 2c). This active site pocket is highly conserved on the conservation surface of MGL proteins, whereas other surface regions have low amino acid conservation (Figure 2d). This assembly in the active dimer might be necessary for the catalytic reaction. The Fn1419 tetrameric structure is mainly stabilized by hydrogen-bond linkages at the dimer–dimer interface and functions to maintain an active dimer.

Next, we compared the crystal structure of Fn1419 with those of PLP-dependent MGLs and CGLs. Fn1419 is highly structurally similar to representative PLP-dependent MGL proteins from *Pseudomonas putida* (PpMGL; PDB entry: 5X2W; sequence identity = 54%; r.m.s. deviation = 0.81 Å), *Citrobacter freundii* (CfMGL; 6S0C; 58%; 0.56 Å), and *Clostridium sporogenes* (CsMGL; 5DX5; 67%; 0.51 Å), and also to PLP-dependent CGL from Bacillus cereus (BcCGL; 7D7O; 49%; 0.80 Å), *Staphylococcus aureus* (SaCGL; 6KGZ; 42%; 0.90 Å), and *Pseudomonas aeruginosa* (PaCGL; 7BA4; 44%; 0.73 Å). Superimposition of the MGL and CGL proteins showed large conformation differences in PBD. In particular, the loop region between the α2- and α3-helix of the crystal structure of Fn1419, CfMGL, SaCGL, and PaCGL are disordered (Figure 3a,b), which indicates that the α2-α3 loop is highly flexible in the MGL proteins family.

2.3. Substrate- and Cofactor-Binding Sites of Fn1419

In MGLs, PLP- and substrate-binding sites are located between the PBD and CTD [22–24]. Although the PLP-bound Fn1419 solution with the yellowish color was used for the crystallization, the electron density map corresponding to the PLP molecule was not observed in the predicted PLP-binding site. This observation indicates that the crystal structure of Fn1419 is in a PLP-free state, and PLP molecules were dissociated from Fn1419 during protein crystallization.

The crystal structure of this PLP-free state of Fn1419 was compared with the crystal structure of Fn1419 complexed with LLP, a covalent complex of lysine and PLP (Fn1419-6LXU, unpublished). In Fn1419-6LXU, the PLP-binding pocket is located above the positively charged pocket and is formed by five residues (Gly86, Met87, Ser206, Thr208, and Gly213) (Figure 3a). Further, in Fn1419-6LXU, the phosphate group of LLP is coordinated by four hydrogen bonds, three of which are bifurcated. The OP1 atom of PLP has interacted with the main chains of Gly86 and Ser206 and the side chain of Thr208 by hydrogen bonds, while the OP3 atom of PLP accepts a hydrogen bond from the amide group of Met87. The lysine part of LLP has interacted with Ser206, Thr208, and Gly213 via two water molecules. In addition, a structural comparison between Fn1419 and Fn1419-6LXU reveals no significant conformational change in the residues surrounding the active site (Figure 3a).

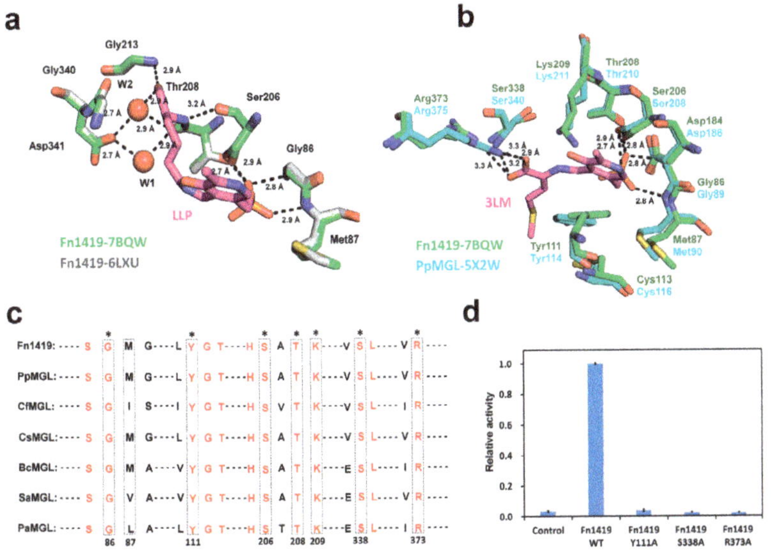

Figure 3. Analysis of PLP- and substrate-binding sites on Fn1419. (**a**) Superimposition of the active sites of PLP-free (PDB code: 7BQW) and PLP-bound (6LXU) states of Fn1419. (**b**) Superimposition of the active sites of Fn1419 and PpMGL-Met (5X2W). (**c**) Partial sequence alignment of MGL enzymes from seven bacteria [*Pseudomonas putida* (Uniprot accession no. P13254), *Citrobacter freundii* (Q84AR1), *Clostridium sporogenes* (A0A1J1CYV7), *Bacillus cereus* (Q818A3), *Staphylococcus aureus* (SAV0460), *and P. aeruginosa* (A0A2R3IX25)]. The PLP-binding residues are indicated by dotted boxes, and conserved residues for PLP binding are indicated by asterisks. All conserved residues shown in red. The numbering refers to Fn1419. (**d**) Measurement of the Fn1419 activity with Fn1419 mutants (Y111A, S338A, and R373A). The error bars represent the standard deviation of three replicates.

Next, the crystal structure of Fn1419 determined in this experiment was compared with that of *Pseudomonas putida* MGL (PpMGL) with 3LM, a complex of L-methionine and PLP (PDB entry: 5X2W) [22] (Figure 3b). In the PpMGL structure, the carboxyl group of the L-methionine substrate forms three hydrogen bonds with two conserved Arg375 and Ser340 residues. The O1 atom of the L-methionine substrate accepts a hydrogen bond from the NH1 atom of Arg375. In contrast, the O2 atom interacts with the amino group of Ser340 and the guanidium group of Arg375 via two hydrogen bonds. Further, the amino group of the L-methionine substrate is stabilized by a hydrogen bond with the hydroxyl group of Tyr114. Comparative analysis of amino acid sequences and crystal structures revealed that Gly86, Met87, Tyr111, Ser206, Thr208, and Lys209 of Fn1419-7BQW are located at the same structural positions as the mentioned corresponding residues of Fn1419-6LXU and PpMGL (Figure 3a,b). The side chains of these six basic amino acids are oriented towards the pocket. In addition, amino acid sequence analysis of Fn1419 and other MGL family proteins revealed that Gly86, Tyr111, Ser206, Thr208, Lys209, Ser338, and Arg373 in Fn1419, except Met87, are highly conserved (Figure 3c and Figure S4). To verify the critical residues for MGL^{Cys} activity of Fn1419, we substituted Tyr111, Ser338, and Arg373 residues of Fn1419 with an alanine residue. Y111A, S338A, and R373A mutants completely inhibited the MGL^{Cys} enzyme activity (Figure 3d). These results demonstrated that Tyr111, S338A, and Arg373 of Fn1419 are essential for substrate binding and are involved in the catalytic reaction.

2.4. Screening of Leading Compounds for Fn1419 Inhibition

Inhibition of H_2S production to control the bacterial defense against ROS is a new strategy for antibacterial drug design [6]. Further, natural products and their structural analogs have historically been essential in pharmacotherapy because they are often assumed to be better tolerated and safer to use than synthetic compound molecules. Although they have side effects, including toxicity, allergic reactions, and drug interactions, they have been reported to be used for many products [25].

To provide insights for inhibitor design, we screened 39 natural compounds, including 21 hydrophilic compounds and 18 hydrophobic compounds, as potential Fn1419 inhibitors (Figure 4a,b). Among the natural hydrophilic compounds screened, the MGL^{Cys} activity of Fn1419 was significantly inhibited by gallic acid (product reduction rate: 84.8%) and slightly reduced by potassium sorbate (5.3%), N-acetyl-cysteine (12%), and creatine monohydrate (11.2%). In contrast, the MGL^{Cys} activity of Fn1419 was enhanced by betaine-HCl (product increase rate: 24%), erythritol (16.4%), choline bitartrate (13.3%), GABA (9.9%), chondroitin sulfate (5.8%), and matrine (4.9%). L-glutamate, chlorogenic acid, citicoline sodium, cordycepin, vitamin C, niacin, inositol, xylitol, L-carnitine, phenylpiracetam, and pyridoxine-HCl did not significantly affect enzyme activity. Considering the hydrophobic compounds, dihydromyricetin (product reduced by 89.5%) significantly inhibited the MGL^{Cys} activity of Fn1419 rutin (product reduction rate: 18.2%); synephrine (32.4%), kaempferol (40.4%), and curcumin (53.9%) mildly inhibited the enzyme. In contrast, the MGL^{Cys} enzyme activity was enhanced by D-raffinose pentahydrate (product increase rate: 72.5%), L-tetrahydropal matrine (71.7%), phenibut HCl (61.8%), 5-hydroxyl tryptophan (53.0%), D-quinic acid (35.1%), hordenine HCl (32.1%), oleamide (24.6%), and noopept (21.8%). It was not significantly affected by α-lipoic acid, indole-3-carbinol, p-coumaric, acetyl L-carnitine HCl, and apigenin. Next, we checked the effects of different concentrations of gallic acid and dihydromyricetin, the two most inhibitory compounds identified, on enzyme activity (Figure 4c,d). To determine the two compounds' inhibitory effects, we performed a titration test to the calculated 50% inhibitory concentration (IC_{50}) values. As a result, the IC_{50} of gallic acid and dihydromyricetin, which had relatively high Fn1419-inhibitory activity compared to other compounds, were 121 and 23 μM, respectively, indicating significant values as a lead compound for inhibitor development.

Considering the chemical structure of gallic acid and dihydromyricetin, we proposed that the trihydroxybenzene group might be critical for this inhibitive effect. To verify this, we investigated the effect of gallic acid analogs (pyrogallic acid, methyl gallate, ethyl gallate, and gallic acid trimethyl ether) on Fn1419 (Figure 4e). The analysis confirmed that trihydroxybenzene-containing pyrogallic acid (product reduction rate: 89.8%), methyl gallate (81.5%), and ethyl gallate (81.9%) significantly decreased, whereas gallic acid trimethyl ether did not affect, the MGL^{Cys} activity of Fn1419 (Figure 4f). The calculated IC_{50} values for dihydromyricetin, gallic acid, pyrogallic acid, and ethyl gallate were 23, 121, 57, and 517 μM, respectively (Figure 4g). These observations indicate that the trihydroxybenzene group is critical for the inhibition of MGL^{Cys} activity of Fn1419, and trihydroxybenzene-containing chemicals are good candidates for Fn1419-targeted antibacterial drug design against *F. nucleatum*.

2.5. Identification of Fn1419-Binding Site for Trihydroxybenzene-Based Lead Compounds

To analyze the trihydroxybenzene-binding mode, we performed the docking of the gallic acid molecule in the crystal structure of Fn1419-7BQW. The analysis suggested that gallic acid forms hydrogen bonds with Gly112, Lys209, Val337, and Arg373, and that the trihydroxybenzene ring is stabilized by Pro159 via a π–π interaction (Figure 4h). The gallic acid-binding site is close to the PLP-binding site on Fn1419, but only one residue from the PLP-binding site (Lys209) is involved in the catalytic activity (Figure 4h). We also modeled docking of pyrogallic acid, methyl gallate, ethyl gallate, and gallic acid trimethyl ether with Fn1419-7BQW, which showed all these compounds were bound to the same position by the trihydroxybenzene group (Figure S5).

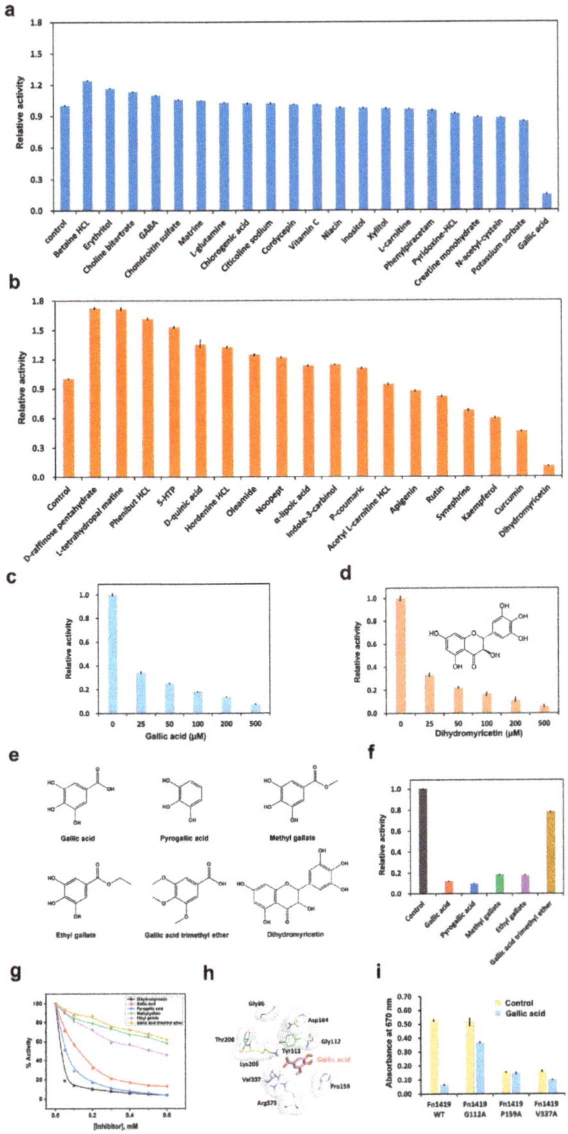

Figure 4. Screening for natural inhibitors of Fn1419. Inhibitor screening of (**a**) 21 hydrophilic compounds and (**b**) 18 hydrophobic compounds to Fn1419 in vitro. The error bars represent standard deviation of three replicates. Titration of (**c**) gallic acid and (**d**) dihydromyricetin to Fn1419. The error bars represent the standard deviation of three replicates. (**e**) Chemical structures of gallic acid, pyrogallic acid, methyl gallate, ethyl gallate, gallic acid trimethyl ether, and dihydromyricetin. (**f**) Fn1419 inhibition study with pyrogallic acid, methyl gallate, ethyl gallate, and gallic acid trimethyl ether. The error bars represent standard deviation of three replicates. (**g**) IC_{50} values of Fn1419 for dihydromyricetin, gallic acid, pyrogallic acid, methyl gallate, ethyl gallate, and gallic acid trimethyl ether. (**h**) Docking of gallic acid to Fn1419. Gallic acid or PLP-binding residue are indicated by gray and green sticks, respectively. (**i**) Measurement of Fn1419 activity with Fn1419 mutants (G112A, P159A, and V337A) predicted the gallic acid binding. The error bars represent standard deviation of three replicates.

To confirm this molecular docking result, we generated Fn1419 variants with Gly112, Pro159, and Val337 substituted with an alanine residue and assayed their MGL^{Cys} activity with and without gallic acid (Figure 4i). Fn1419-G112A retained its enzymatic activity, indicating the inhibitory effect of gallic acid was not pronounced. By contrast, Fn1419-P159A and Fn1419-V337A lost enzyme activity without added gallic acid. These observations suggest that residues P159 and V337 of Fn1419 are related to substrate binding while Gly112 of Fn1419 is critical for gallic acid binding. Indeed, the molecular docking analysis indicated that the gallic acid-binding site of Fn1419 is close to the PLP-binding cavity (Figure 4h).

The inhibitory effect of gallic acid against other PLP-dependent H_2S-producing enzymes of *F. nucleatum* was also tested, using Fn1220, Fn1055, and Fn0625 (Figure 5a). The activity against Fn0625 was too low (H_2S productive rate is only 11.6% compared to Fn1419) to detect inhibition, but gallic acid did not inhibit Fn1220 and Fn1055. These results suggest that gallic acid selectively inhibits the MGL^{Cys} of Fn1419.

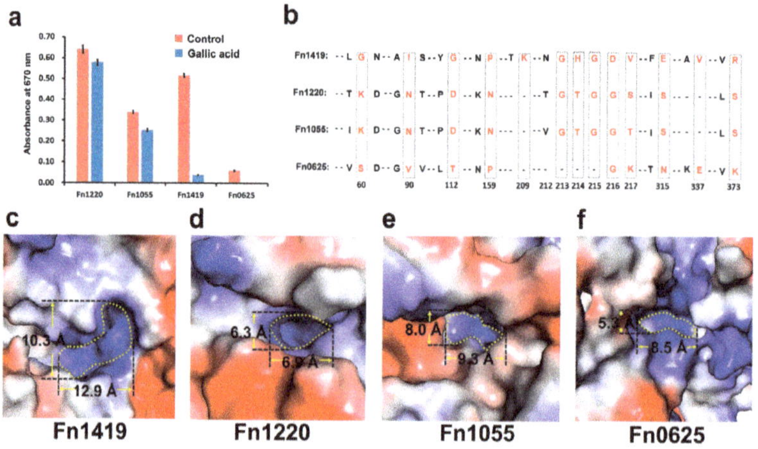

Figure 5. Putative gallic acid-binding sites on Fn1419. (**a**) The inhibitory effect of gallic acid on Fn1220, Fn1055, Fn1419, and Fn0625. The error bars represent the standard deviation of three replicates. (**b**) Partial sequence alignment of the active sites of Fn1220, Fn1055, Fn1419, and Fn0625. The numbering reflects Fn1419 numbering. The identified PLP- and gallic acid-binding residues are boxed and shown in red. (**c**–**f**) Electrostatistic surface of the PLP- and substrate-binding pockets of Fn1419 (PDB ID: 7BQW), Fn1220 (5XEO), Fn1055 (5B53), and Fn0625 (predicted by AlphaFold).

Next, to understand this selective inhibitory activity, the amino acid sequences and crystal structures of Fn1220 and Fn1055 were analyzed (Figure 5b and Figure S6). Sequence alignment revealed a low overall sequence similarity with Fn1419: 11% for Fn1220 and 13% for Fn1055. Of note, although the amino acid similarity in the PLP-binding domain was high, Fn1419 residues that are important for gallic acid binding (Gly112, Pro159, and Val337) were not conserved (Figure 5b). Next, the PLP- and substrate-binding pocket of Fn1419 were compared with those of Fn1220, Fn1055, and Fn0625. The PLP-binding pocket is wider than that of other proteins (Figure 5c–f), which indicates that Fn1419 is able to bind to gallic acid, unlike Fn1220, Fn1055, and Fn0625. Based on the above, we propose that the differences in the PLP-binding pocket may be the reason for the selective inhibition of Fn1419 by gallic acid. Further studies on the crystal structure of Fn1419 complexed with gallic acid are required to clarify the utility of the PLP-binding pocket as a target of drug design.

3. Discussion

Bacteria produce endogenous H_2S as a defense against ROS and antibiotic-induced oxidative damage [26–28], and H_2S biogenesis is a critical contributor to bacterial antibiotic tolerance and a target for versatile antibiotic potentiators [6]. In *F. nucleatum*, the MGL^{Cys} protein Fn1419 is involved in synthesizing H_2S, together with three other enzymes (Fn1220, Fn1055, and Fn0625) [16], but there are significant differences between these enzymes [16]. The crystal structures of Fn1220 and Fn1055 have been determined by Yuichiro Kezuka et al. [29]. However, until now, the crystal structures and molecular functions of Fn1419 and Fn0625 remained largely uncharacterized. Based on these, we presented the first structural and functional characterization of Fn1419 and screened for lead compound from many natural products. As a result, we found that two natural chemicals, hydrophilic gallic acid and hydrophobic dihydromyricetin, can selectively inhibit the enzyme activity of Fn1419. In addition, we also analyzed the structural insight of Fn1419 by molecule docking to understand the interaction mechanism of gallic acid with Fn1419. The crystal structure and the discovery of lead compound in Fn1419, provide a foundation for the development of selective inhibitors for this poorly studied gene.

What is noteworthy is that we obtained the crystal structure of Fn1419. The overall structure shared structure organization with its homologues (Figure S3a,b). A structural comparison of the apo- and PLP/substrate-bound structures of Fn1419 showed that the key amino acids at the active center for cofactor/substrate binding share substantially the same conformational constraints, which suggests our structure has the ability to bind PLP and its substrates. In addition, mutant test analyses of Fn1419 suggest that Tyr111, Ser338, and Arg373 are very important for PLP and substrate binding, which lose degradation activity toward cysteine after the mutation to Ala. We guess all that because the mutation disrupts hydrogen bonds and the active center.

Many bactericidal antibiotics kill the bacteria by stimulating the production of highly toxic hydroxyl radicals, whose production is mediated by Fe^{2+} [30]. Free cysteine could accelerate the Fe^{2+}-mediated Fenton reaction by reducing Fe^{3+} to Fe^{2+} [31]. To deal with the stress, the bacteria could alter gene expression levels and metabolism, which are possibly linked to inhibition of the production of hydroxyl radicals [32], which could be an inhibitor target to screen for specific inhibitors [33]. In the study, we found that gallic acid and dihydromyricetin can significantly inhibit Fn1419 enzyme activity. Considering dihydromyricetin has low solubility and permeability, as well as poor bioavailability [34], experiments mainly focused on the gallic acid, which has many biological properties, including antimicrobial, anticancer, antioxidant, and anti-inflammatory [35]. It strongly inhibits the growth of various bacteria, including antibiotic-resistant *S. aureus*, *E. coli*, *Mannheimia haemolytica*, and *Pasteurella multocida* strains [36], by inhibiting efflux pumps and folate metabolism. Additionally, gallic acid can potentiate the efficacy of antimicrobials [37]. To verify the important role of trihydroxybenzene, we selected four gallic acid analogues, containing trihydroxybenzene or not, to detect the inhibition. This result was expected since the natural compounds containing trihydroxybenzene have an inhibitory effect on H_2S-producing enzymes from anaerobic pathogenic bacteria, which has not been reported. Due to the low IC_{50} value of gallic acid and its analogs for Fn1419, these chemicals are not suitable for application to completely inhibit Fn1419. However, we believe that gallic acid trihydroxybenzene can be used as a starting point to design and synthesize more effective inhibitors in the future.

On the other hand, gallic acid did not inhibit the H_2S production of Fn1220 and Fn1055. To investigate the structural characteristics of these proteins that lead to the differences in the inhibitory efficacy of gallic acid on the structures and sequences of Fn1419, we analyzed Fn1220 and Fn1055. Fn1220 and Fn1055 lack key amino acid residues involved in gallic acid binding to Fn1419 (Gly112, Pro159, and Val337 on Fn1419). Further, using molecular docking and activity assays with enzyme variants, Gly112 was identified as critical to gallic acid binding to Fn1419. However, crystal structures of the inhibitor-bound complexes should be determined to delineate the role of different amino acids of Fn1419 in its sensitivity to different inhibitors.

In summary, we found that two natural compounds, gallic acid and dihydromyricetin, selectively inhibit Fn1419. Furthermore, analysis of the inhibitory effect of gallic acid analogs on Fn1419 revealed that the trihydroxybenzene component of the molecule is potentially responsible for the observed inhibition of enzymatic activity. These observations constitute a starting point for developing more potent and selective inhibitors that could also serve as a powerful tool to clarify the biological roles of this H_2S-producing enzyme from *F. nucleatum*.

4. Materials and Methods

4.1. Protein Preparation

The sequence encoding Fn1419 (GeneBank: AAL95612.1, residues 1–395) was obtained by PCR amplification, using the genomic DNA of *F. nucleatum* ATCC 25586 as a template. The gene was cloned into the *Nco*I and *Xho*I sites of the PET-28a (+) vector (Invitrogen, Carlsbad, CA, USA) to express a recombinant protein with his-tag at the N-terminal. The resultant plasmid was transformed into *E. coli* BL21 (DE3) cells. Transformed bacteria were grown in 1 L of LB medium containing 50 mg/mL kanamycin at 37 °C until OD_{600} of 0.6–0.8. Recombinant protein expression was induced by adding 0.5 mM isopropyl-β-D-thiogalactoside (IPTG) at 30 °C for 8 h. The cells were harvested by centrifugation and resuspended in lysis buffer (20 mM Tris-HCl, pH 8.0, 150 mM NaCl, and 2 mM β-mercaptoethanol). The cells were lysed by sonication; then, centrifugation at 13,000 rpm for 30 min at 4 °C was performed to remove the cell debris. The supernatant was preincubated with Ni^{2+}-NTA affinity resin (GE Healthcare, Chicago, IL, USA) for 20 min at 4 °C and then loaded onto a column. The resin was washed with washing buffer (20 mM Tris-HCl, pH 8.0, 150 mM NaCl, 2 mM β-mercaptoethanol, and 30 mM imidazole). The recombinant protein was eluted with elution buffer (20 mM Tris-HCl, pH 8.0, 150 mM NaCl, 2 mM β-mercaptoethanol, and 250 mM imidazole). The protein was concentrated by a Vivaspin centrifugal concentrator MWCO 30 kDa (Vivaspin, Littleton, MA, USA) and loaded onto a pre-equilibrated HiLoad Superdex 200 26/60 column (GE Healthcare, Chicago, IL, USA) with pre-equilibrated buffer containing 20 mM Tris-HCl, pH 8.0, 150 mM NaCl, and 2 mM β-mercaptoethanol. The protein was concentrated to 20 mg/mL using a Vivaspin centrifugal concentrator MWCO 30 kDa (Vivaspin, Littleton, MA, USA) and stored at −80 °C. The purity of the purified protein was confirmed by 15% sodium dodecyl sulfate-polyacrylamide gel electrophoresis (SDS-PAGE) and Coomassie blue R-250 staining. The sequence encoding Fn1220 (GeneBank: AAL95416.1, residues 1–316), Fn1055 (GeneBank: AAL95251.1, residues 1–336), and Fn0625 (GeneBank: AAL94821.1, residues 1–398) was constructed, expressed, and purified using the same procedure as that used for wild-type Fn1419. Site-directed mutagenesis was performed by using two subsequent PCR reactions [38], and protein variants were expressed and purified, also using the same procedure as that used for wild-type Fn1419. The oligonucleotide sequences used for PCR reactions are listed in Table S2.

4.2. Analytical Size-Exclusion Chromatography

Purified Fn1419 (500 μL; 5 mg/mL) was loaded onto a Superdex 200 10/300 GL column (GE Healthcare, USA), and the protein was eluted using a buffer containing 20 mM Tris-HCl, pH 8.0, 150 mM NaCl, and 2 mM β-mercaptoethanol, at a flow rate of 0.3 mg/mL. The molecular weights of the eluted samples were calculated based on the calibration curves by β-amylase and alcohol dehydrogenase (Gel Filtration Markers Kit for Protein Molecular Weights 12,000–200,000 Da, Sigma-Aldrich, St. Louis, MO, USA).

4.3. PLP Binding Analysis

Absorptions of Fn1419 (0.18 mM) in buffer containing 20 mM Tris-HCl, pH 8.0, 150 mM NaCl, and 2 mM β-mercaptoethanol, and Pyridoxal 5 -phosphate (PLP, P9255, Sigma-Aldrich) were measured using ultraviolet-visible (UV-Vis-2450) spectroscopy (Shimadzu Corporation) at 25 °C in the wavelength range of 250–600 nm.

4.4. Enzyme Assays

MGL^{Cys} activity of Fn1419 was measured as previously reported [16]. Fn1419 (5.75 nM) in assay buffer (40 mM potassium phosphate, pH 7.6, and 10 μM PLP) was mixed with various concentrations of L-cysteine (25, 50, 100, 200, 400, 800, or 1000 μM) and incubated for 15 min at 37 °C. H_2S, the reaction product, was then detected based on methylene blue formation [16]. Specifically, 100 μL of solution A (the enzyme reaction solution), 20 μL of solution B (20 mM N′,N′-dimethyl-p-phenylenediamine dihydrochloride and 7.2 M HCl), and 20 μL of solution C (30 mM $FeCl_3$ and 1.2 M HCl) were mixed, and the mixture incubated for 30 min at room temperature. The generated methylene blue exhibited maximum absorbance at 670 nm and was detected using a SpectraMax 190 Microplate Reader (Molecular Devices, San Jose, CA, USA) at 25 °C.

4.5. Crystallization, Data Collection, and Structure Determination

Fn1419 solution was concentrated to 20 mg/mL using a Vivaspin centrifugal concentrator MWCO 30 kDa (Vivaspin, Littleton, MA, USA). The initial crystallization was performed using the sitting-drop vapor-diffusion method at 22 °C with commercially available crystallization screening kits, namely, PEGRx (Hampton Research, Aliso Viejo, CA, USA). Each crystal drop was composed of 0.5 μL of protein solution (20 mM Tris-HCl, pH 8.0, 150 mM NaCl, and 2 mM β-mercaptoethanol) and 0.5 μL of reservoir solution. Finally, an Fn1419 crystal was grown in a precipitate composed of 20% (*w*/*v*) polyethylene glycol 5000, 100 mM bismuth (pH 8.5), and 200 mM sodium formate at 22 °C within 5 days. The crystal was removed from the drop and immediately flash-cooled in liquid nitrogen at −173 °C.

X-ray diffraction data of Fn1419 were collected at beamline BL19U1 at the Shanghai Synchrotron Radiation Facility (SSRF) under a nitrogen stream at −173 °C. The X-ray wavelength was 0.9792 Å. A total of 360 diffraction images were collected. The diffraction data were indexed, integrated, and scaled using the HKL2000 program [39]. Initial phases were obtained using the molecular replacement method, using the Phaser [40] with the crystal structure of CsMGL (PDB entry 5DX5) as a search model [24]. Manual model-building was performed using the COOT program [41] and refined using phenix.refinement in the PHENIX package [42]. The data collection and refinement statistics are shown in Table 1. The atomic coordinates and structure factors were deposited in the Protein Data Bank (PDB entry 7BQW). All structural figures were drawn using PyMOL [43]. The tetrameric interface of Fn1419 was calculated using a PISA server available from EBI-EMBL [44]. The surface conservation of Fn1419 was calculated using the Consurf server [45]. The sequence alignment was generated using the Clustal Omega server [46]. The model structure of Fn0625 was generated by Alphafold [47].

4.6. Inhibitor Screening Assay

Fn1419 (5.75 nM) in assay buffer (40 mM potassium phosphate, pH 7.6 and 10 μM PLP) was mixed with 100 μM of different classes of 21 hydrophilic compounds (betaine-HCl, erythritol, choline bitartrate, GABA, chondroitin sulfate, matrine, L-glutamine, chlorogenic acid, citicoline sodium, cordycepin, vitamin C, niacin, inositol, xylitol, L-carnitine, phenylpiracetam, pyridoxine-HCl, creatine monohydrate, N-acetyl-cysteine, potassium sorbate, and gallic acid) and 18 hydrophobic compounds (d-raffinose pentahydrate, L-tetrahydropal matrine, phenibut-HCl, 5-HTP, d-quinic acid, hordenine-HCl, oleamide, noopept, α-lipoic acid, indole-3-carbinol, p-coumaric acid, acetyl L-carnitine-HCl, apigenin, futin, synephrine, kaempferol, curcumin, and dihydromyricetin) and incubated for 20 min at 20 °C. Then, 500 μM L-cysteine was added, and the reaction was allowed to proceed for 5 min at 37 °C. The amount of produced H_2S was detected indirectly, based on the formation of methylene blue [16]. The reaction mixture consisted of solution A (100 μL), solution B (20 μL), and solution C (20 μL) (the solution composition is described at the enzyme assays) and was incubated for 30 min at 20 °C. The amount of generated methylene blue was detected at 670 nm, using a SpectraMax 190 Microplate Reader (Molecular

Devices, San Jose, CA, USA) at 25 °C. The inhibitory effect of four compounds (pyrogallic acid, methyl gallate, ethyl gallate, and gallic acid trimethyl ether) was tested using the same procedure as that used for the natural compounds. The inhibitory effects of gallic acid on Fn1220 and Fn1055 were tested using the same method as above. The concentration of both enzymes was 5.75 nM, and the concentration of gallic acid was 500 μM.

4.7. IC_{50} Determination

Based on the changes in the MGL^{Cys} activity of Fn1419, after treatment with dihydromyricetin, gallic acid, pyrogallic acid, methyl gallate, ethyl gallate, and gallic acid trimethyl ether for 20 min at 20 °C, the generation of methylene blue was detected at 670 nm using a SpectraMax 190 Microplate Reader (Molecular Devices, San Jose, CA, USA) at 25 °C. The concentration of the enzyme was 5.75 nM and the concentration of the six above-mentioned compounds was 500 μM. IC_{50} values were eventually calculated from non-linear regression curves fitted.

4.8. Molecular Docking

Molecular docking of gallic acid, pyrogallic acid, and dihydromyricetin in Fn1419 was performed using the High Ambiguity Driven Protein-Protein Docking (HADDOCK) web server [48]. The 3D structures of the three small molecules (gallic acid, pyrogallic acid, and dihydromyricetin) were obtained from the PDB. Water molecules in the crystal structure of Fn1419 (PDB: 7BQW) were removed before molecular docking. The residues (81–125, 205–225, 335–355, 372–385) were defined based on the position of the active site cavity, PLP, substrate-binding residues, and the structural characteristics of small molecules. The optimal pose was selected for further study based on the HADDOCK score.

Supplementary Materials: The following supporting information can be downloaded at https://www.mdpi.com/article/10.3390/ijms24021651/s1.

Author Contributions: K.H.N. and Y.X. Conceptualization; T.B., J.L., C.Q., K.H.N. and Y.X. methodology; T.B., J.L., J.Z., X.B., S.H., L.W., Y.J. and X.J. (Xiaoyu Jin) formal analysis; T.B., J.L., J.Z., X.B., S.H., X.J. (Xiaoling Jin), L.Z., K.H.N. and Y.X., supervision; C.Q., K.H.N. and Y.X. validation; T.B., J.L., I.J., L.Z., K.H.N. and Y.X. investigation; T.B., J.L., K.H.N. and Y.X. writing review and editing; H.P., N.-C.H., C.Q., K.H.N. and Y.X. resources; K.H.N. and Y.X. funding acquisition. All authors have read and agreed to the published version of the manuscript.

Funding: This work was supported by the Natural Science Foundation of Liaoning Province (grant no. 2019-MS-065 to Y.X.) and the National Research Foundation of Korea (grant Nos. NRF-2017M3A9F6029736; NRF-2020M3H1A1075314; NRF-2021R1I1A1A01050838 to K.H.N.). This study was supported by ProGen (to K.H.N.).

Institutional Review Board Statement: Not applicable.

Informed Consent Statement: Not applicable.

Data Availability Statement: The coordinates and structure factor of Fn1419 deposited on PDB (PDB entry 7BQW). All data described in the manuscript are contained in the manuscript and the associated Supporting Information files.

Acknowledgments: The authors acknowledge the staff on beamline BL19U1 of the SSRF for their assistance with data collection.

Conflicts of Interest: The authors declare no conflict of interest.

References

1. Brook, I. Spectrum and treatment of anaerobic infections. *J. Infect. Chemother.* **2016**, *22*, 1–13. [CrossRef] [PubMed]
2. Hecht, D.W. Prevalence of antibiotic resistance in anaerobic bacteria: Worrisome developments. *Clin. Infect. Dis.* **2004**, *39*, 92–97. [CrossRef] [PubMed]
3. Li, H.; Zhou, X.; Huang, Y.; Liao, B.; Cheng, L.; Ren, B. Reactive Oxygen Species in Pathogen Clearance: The Killing Mechanisms, the Adaption Response, and the Side Effects. *Front. Microbiol.* **2020**, *11*, 622534. [CrossRef] [PubMed]

. Xiao, Q.; Ying, J.; Xiang, L.; Zhang, C. The biologic effect of hydrogen sulfide and its function in various diseases. *Medicine* **2018**, *97*, e13065. [CrossRef] [PubMed]

Tabassum, R.; Jeong, N.Y.; Jung, J. Protective effect of hydrogen sulfide on oxidative stress-induced neurodegenerative diseases. *Neural Regen. Res.* **2020**, *15*, 232–241.

Shatalin, K.; Nuthanakanti, A.; Kaushik, A.; Shishov, D.; Peselis, A.; Shamovsky, I.; Pani, B.; Lechpammer, M.; Vasilyev, N.; Shatalina, E.; et al. Inhibitors of bacterial H_2S biogenesis targeting antibiotic resistance and tolerance. *Science* **2021**, *372*, 1169–1175. [CrossRef]

. Han, Y.W. *Fusobacterium nucleatum*: A commensal-turned pathogen. *Curr. Opin. Microbiol.* **2015**, *23*, 141–147. [CrossRef]

. Sun, C.H.; Li, B.B.; Wang, B.; Zhao, J.; Zhang, X.Y.; Li, T.T.; Li, W.B.; Tang, D.; Qiu, M.J.; Wang, X.C.; et al. The role of *Fusobacterium nucleatum* in colorectal cancer: From carcinogenesis to clinical management. *Chronic Dis. Transl. Med.* **2019**, *5*, 178–187. [CrossRef]

. Blachier, F.; Beaumont, M.; Kim, E. Cysteine-derived hydrogen sulfide and gut health: A matter of endogenous or bacterial origin. *Curr. Opin. Clin. Nutr. Metab. Care* **2019**, *22*, 68–75. [CrossRef]

0. Kabil, O.; Vitvitsky, V.; Xie, P.; Banerjee, R. The quantitative significance of the transsulfuration enzymes for H_2S production in murine tissues. *Antioxid. Redox Signal.* **2011**, *15*, 363–372. [CrossRef]

1. Paul, B.D.; Snyder, S.H. H_2S signalling through protein sulfhydration and beyond. *Nat. Rev. Mol. Cell Bio.* **2012**, *13*, 499–507. [CrossRef]

2. Zhao, Y.; Cerda, M.M.; Pluth, M.D. Fluorogenic hydrogen sulfide (H_2S) donors based on sulfenyl thiocarbonates enable H_2S tracking and quantification. *Chem. Sci.* **2019**, *10*, 1873–1878. [CrossRef]

3. Kabil, O.; Banerjee, R. Enzymology of H_2S biogenesis, decay and signaling. *Antioxid. Redox Signal.* **2014**, *20*, 770–782. [CrossRef]

4. Castellarin, M.; Warren, R.L.; Freeman, J.D.; Dreolini, L.; Krzywinski, M.; Strauss, J.; Barnes, R.; Watson, P.; Allen-Vercoe, E.; Moore, R.A.; et al. *Fusobacterium nucleatum* infection is prevalent in human colorectal carcinoma. *Genome Res.* **2012**, *22*, 299–306. [CrossRef]

5. Braccia, D.J.; Jiang, X.; Pop, M.; Hall, A.B. The Capacity to Produce Hydrogen Sulfide (H_2S) via Cysteine Degradation Is Ubiquitous in the Human Gut Microbiome. *Front. Microbiol.* **2021**, *12*, 705583. [CrossRef]

6. Suwabe, K.; Yoshida, Y.; Nagano, K.; Yoshimura, F. Identification of an L-methionine gamma-lyase involved in the production of hydrogen sulfide from L-cysteine in *Fusobacterium nucleatum subsp. nucleatum ATCC 25586. Microbiology* **2011**, *157 Pt 10*, 2992–3000. [CrossRef]

7. Mothersole, R.G.; Wolthers, K.R. Structural and Kinetic Insight into the Biosynthesis of H_2S and l-Lanthionine from l-Cysteine by a Pyridoxal l-Phosphate-Dependent Enzyme from *Fusobacterium nucleatum. Biochemistry* **2019**, *58*, 3592–3603. [CrossRef]

8. Sato, D.; Yamagata, W.; Harada, S.; Nozaki, T. Kinetic characterization of methionine gamma-lyases from the enteric protozoan parasite Entamoeba histolytica against physiological substrates and trifluoromethionine, a promising lead compound against amoebiasis. *FEBS J.* **2008**, *275*, 548–560. [CrossRef]

9. Li, Z.; Zhao, Y.; Zhou, H.; Luo, H.B.; Zhan, C.G. Catalytic Roles of Coenzyme Pyridoxal-5′-phosphate (PLP) in PLP-dependent Enzymes: Reaction Pathway for Methionine-gamma-lyase-catalyzed L-methionine Depletion. *ACS Catal.* **2020**, *10*, 2198–2210. [CrossRef]

20. Toney, M.D. Reaction specificity in pyridoxal phosphate enzymes. *Arch. Biochem. Biophys.* **2005**, *433*, 279–287. [CrossRef]

21. He, S.; Chen, Y.; Wang, L.; Bai, X.; Bu, T.; Zhang, J.; Lu, M.; Ha, N.C.; Quan, C.; Nam, K.H.; et al. Structural and Functional Analysis of the Pyridoxal Phosphate Homeostasis Protein YggS from *Fusobacterium nucleatum. Molecules* **2022**, *27*, 4781. [CrossRef] [PubMed]

22. Sato, D.; Shiba, T.; Yunoto, S.; Furutani, K.; Fukumoto, M.; Kudou, D.; Tamura, T.; Inagaki, K.; Harada, S. Structural and mechanistic insights into homocysteine degradation by a mutant of methionine gamma-lyase based on substrate-assisted catalysis. *Protein Sci.* **2017**, *26*, 1224–1230. [CrossRef] [PubMed]

23. Revtovich, S.V.; Faleev, N.G.; Morozova, E.A.; Anufrieva, N.V.; Nikulin, A.D.; Demidkina, T.V. Crystal structure of the external aldimine of *Citrobacter freundii* methionine gamma-lyase with glycine provides insight in mechanisms of two stages of physiological reaction and isotope exchange of alpha- and beta-protons of competitive inhibitors. *Biochimie* **2014**, *101*, 161–167. [CrossRef]

24. Revtovich, S.; Anufrieva, N.; Morozova, E.; Kulikova, V.; Nikulin, A.; Demidkina, T. Structure of methionine gamma-lyase from *Clostridium sporogenes. Acta Crystallogr. F Struct. Biol. Commun.* **2016**, *72 Pt 1*, 65–71. [CrossRef] [PubMed]

25. Atanasov, A.G.; Zotchev, S.B.; Dirsch, V.M.; International Natural Product Sciences, T.; Supuran, C.T. Natural products in drug discovery: Advances and opportunities. *Nat. Rev. Drug Discov.* **2021**, *20*, 200–216. [CrossRef] [PubMed]

26. Mironov, A.; Seregina, T.; Nagornykh, M.; Luhachack, L.G.; Korolkova, N.; Lopes, L.E.; Kotova, V.; Zavilgelsky, G.; Shakulov, R.; Shatalin, K.; et al. Mechanism of H_2S-mediated protection against oxidative stress in *Escherichia coli. Proc. Natl. Acad. Sci. USA* **2017**, *114*, 6022–6027. [CrossRef]

27. Toliver-Kinsky, T.; Cui, W.; Toro, G.; Lee, S.J.; Shatalin, K.; Nudler, E.; Szabo, C. H_2S, a Bacterial Defense Mechanism against the Host Immune Response. *Infect. Immun.* **2019**, *87*, e00272-18. [CrossRef]

28. Mendes, S.S.; Miranda, V.; Saraiva, L.M. Hydrogen Sulfide and Carbon Monoxide Tolerance in Bacteria. *Antioxidants* **2021**, *10*, 729. [CrossRef]

29. Yoshida, Y.; Suwabe, K.; Nagano, K.; Kezuka, Y.; Kato, H.; Yoshimura, F. Identification and enzymic analysis of a novel protein associated with production of hydrogen sulfide and L-serine from L-cysteine in *Fusobacterium nucleatum subsp. nucleatum ATCC 25586*. *Microbiology* **2011**, *157 Pt 7*, 2164–2171. [CrossRef]
30. Kohanski, M.A.; Dwyer, D.J.; Hayete, B.; Lawrence, C.A.; Collins, J.J. A common mechanism of cellular death induced by bactericidal antibiotics. *Cell* **2007**, *130*, 797–810. [CrossRef]
31. Park, S.; Imlay, J.A. High levels of intracellular cysteine promote oxidative DNA damage by driving the fenton reaction. *J Bacteriol.* **2003**, *185*, 1942–1950. [CrossRef]
32. Somerville, G.A.; Proctor, R.A. At the crossroads of bacterial metabolism and virulence factor synthesis in *Staphylococci*. *Microbiol. Mol. Biol. Rev.* **2009**, *73*, 233–248. [CrossRef]
33. Poole, K. Bacterial stress responses as determinants of antimicrobial resistance. *J. Antimicrob. Chemother.* **2012**, *67*, 2069–2089 [CrossRef]
34. Zhang, H.; Caprioli, G.; Hussain, H.; Le, N.P.K.; Farag, M.A.; Xiao, J. A multifaceted review on dihydromyricetin resources, extraction, bioavailability, biotransformation, bioactivities, and food applications with future perspectives to maximize its value *eFood* **2021**, *2*, 164–184. [CrossRef]
35. Yang, K.; Zhang, L.; Liao, P.; Xiao, Z.; Zhang, F.; Sindaye, D.; Xin, Z.; Tan, C.; Deng, J.; Yin, Y.; et al. Impact of Gallic Acid on Gut Health: Focus on the Gut Microbiome, Immune Response, and Mechanisms of Action. *Front. Immunol.* **2020**, *11*, 580208. [CrossRef]
36. Rajamanickam, K.; Yang, J.; Sakharkar, M.K. Gallic Acid Potentiates the Antimicrobial Activity of Tulathromycin against Two Key Bovine Respiratory Disease (BRD) Causing-Pathogens. *Front. Pharmacol.* **2018**, *9*, 1486. [CrossRef]
37. Cota, D.; Patil, D. Antibacterial potential of ellagic acid and gallic acid against IBD bacterial isolates and cytotoxicity against colorectal cancer. *Nat. Prod. Res.* **2022**, *15*, 1–5. [CrossRef]
38. Allemandou, F.; Nussberger, J.; Brunner, H.R.; Brakch, N. Rapid Site-Directed Mutagenesis Using Two-PCR-Generated DNA Fragments Reproducing the Plasmid Template. *J. Biomed. Biotechnol.* **2003**, *2003*, 202–207. [CrossRef]
39. Otwinowski, Z.; Minor, W. Processing of X-ray diffraction data collected in oscillation mode. *Methods Enzymol.* **1997**, *276*, 307–326.
40. Evans, P.; McCoy, A. An introduction to molecular replacement. *Acta Crystallogr. D Biol. Crystallogr.* **2008**, *64 Pt 1*, 1–10. [CrossRef]
41. Emsley, P.; Cowtan, K. Coot: Model-building tools for molecular graphics. *Acta Crystallogr. D* **2004**, *60*, 2126–2132. [CrossRef] [PubMed]
42. Adams, P.D.; Grosse-Kunstleve, R.W.; Hung, L.W.; Ioerger, T.R.; McCoy, A.J.; Moriarty, N.W.; Read, R.J.; Sacchettini, J.C.; Sauter, N.K.; Terwilliger, T.C. PHENIX: Building new software for automated crystallographic structure determination. *Acta Crystallogr. D* **2002**, *58*, 1948–1954. [CrossRef] [PubMed]
43. DeLano, W.L. Use of PYMOL as a communications tool for molecular science. *Abstr. Pap. Am. Chem. Soc.* **2004**, *228*, U313–U314.
44. Krissinel, E.; Henrick, K. Inference of macromolecular assemblies from crystalline state. *J. Mol. Biol.* **2007**, *372*, 774–797. [CrossRef]
45. Ashkenazy, H.; Abadi, S.; Martz, E.; Chay, O.; Mayrose, I.; Pupko, T.; Ben-Tal, N. ConSurf 2016: An improved methodology to estimate and visualize evolutionary conservation in macromolecules. *Nucleic Acids Res.* **2016**, *44*, W344–W350. [CrossRef]
46. Sievers, F.; Wilm, A.; Dineen, D.; Gibson, T.J.; Karplus, K.; Li, W.; Lopez, R.; McWilliam, H.; Remmert, M.; Soding, J.; et al. Fast, scalable generation of high-quality protein multiple sequence alignments using Clustal Omega. *Mol. Syst. Biol.* **2011**, *7*, 539. [CrossRef]
47. Jumper, J.; Evans, R.; Pritzel, A.; Green, T.; Figurnov, M.; Ronneberger, O.; Tunyasuvunakool, K.; Bates, R.; Zidek, A.; Potapenko, A.; et al. Highly accurate protein structure prediction with AlphaFold. *Nature* **2021**, *596*, 583–589. [CrossRef]
48. De Vries, S.J.; van Dijk, M.; Bonvin, A.M.J.J. The HADDOCK web server for data-driven biomolecular docking. *Nat. Protoc.* **2010**, *5*, 883–897. [CrossRef]

International Journal of
Molecular Sciences

MDPI

Article

Structural and Biochemical Analyses of the Butanol Dehydrogenase from *Fusobacterium nucleatum*

Xue Bai [1,2,†], Jing Lan [1,2,†], Shanru He [1,2], Tingting Bu [1,2], Jie Zhang [1,2], Lulu Wang [1,2,3], Xiaoling Jin [1,2], Yuanchao Mao [1,2], Wanting Guan [1,2], Liying Zhang [1,2], Ming Lu [4], Hailong Piao [5], Inseong Jo [6], Chunshan Quan [1,2], Ki Hyun Nam [7,8,*] and Yongbin Xu [1,2,*]

1 Department of Bioengineering, College of Life Science, Dalian Minzu University, Dalian 116600, China
2 Key Laboratory of Biotechnology and Bioresources Utilization of Ministry of Education, College of Life Science, Dalian Minzu University, Dalian 116600, China
3 School of Life Science and Biotechnology, Dalian University of Technology, No. 2 Linggong Road, Dalian 116024, China
4 Shandong Provincial Key Laboratory of Energy Genetics, Key Laboratory of Biofuel, Qingdao Institute of Bioenergy and Bioprocess Technology, Chinese Academy of Sciences, Qingdao 266101, China
5 CAS Key Laboratory of Separation Science for Analytical Chemistry, Dalian Institute of Chemical Physics, Chinese Academy of Sciences, Dalian 116023, China
6 Infectious Diseases Therapeutic Research Center, Korea Research Institute of Chemical Technology, Daejeon 34114, Republic of Korea
7 Department of Life Science, Pohang University of Science and Technology, Pohang 35398, Republic of Korea
8 POSTECH Biotech Center, Pohang University of Science and Technology, Pohang 35398, Republic of Korea
* Correspondence: structures@postech.ac.kr (K.H.N.); yongbinxu@dlnu.edu.cn (Y.X.)
† These authors contributed equally to this work.

Abstract: Butanol dehydrogenase (BDH) plays a significant role in the biosynthesis of butanol in bacteria by catalyzing butanal conversion to butanol at the expense of the NAD(P)H cofactor. BDH is an attractive enzyme for industrial application in butanol production; however, its molecular function remains largely uncharacterized. In this study, we found that *Fusobacterium nucleatum* YqdH (FnYqdH) converts aldehyde into alcohol by utilizing NAD(P)H, with broad substrate specificity toward aldehydes but not alcohols. An in vitro metal ion substitution experiment showed that FnYqdH has higher enzyme activity in the presence of Co^{2+}. Crystal structures of FnYqdH, in its apo and complexed forms (with NAD and Co^{2+}), were determined at 1.98 and 2.72 Å resolution, respectively. The crystal structure of apo- and cofactor-binding states of FnYqdH showed an open conformation between the nucleotide binding and catalytic domain. Key residues involved in the catalytic and cofactor-binding sites of FnYqdH were identified by mutagenesis and microscale thermophoresis assays. The structural conformation and preferred optimal metal ion of FnYqdH differed from that of TmBDH (homolog protein of FnYqdH). Overall, we proposed an alternative model for putative proton relay in FnYqdH, thereby providing better insight into the molecular function of BDH.

Keywords: *Fusobacterium nucleatum*; butanol dehydrogenase (BDH); YqdH; NAD(P)H; metal ion

1. Introduction

Butyrate biosynthesis by the gut microbiota plays a critical role in maintaining health and, if altered, can lead to pathological conditions, such as inflammatory bowel diseases [1], ulcerative colitis, and type II diabetes [2,3]. Butyrate, the preferred energy source for colonocytes, is protective in colorectal cancer [4]. It is generated from carbohydrates via the glycolysis of two acetyl-CoA molecules yielding acetoacetyl-CoA, followed by stepwise reduction to butyryl-CoA [3]. Four pathways with different initial substrates (pyruvate, 4-aminobutyrate, glutarate, and lysine) are known for butyrate formation [5,6]. Each of these four pathways uses the butyryl-CoA dehydrogenase-electron transferring flavoprotein complex (Bcd-Etf) to catalyze the conversion of crotonyl-CoA to butyryl-CoA [7]. Eventually, the final butyrate production is catalyzed by butyryl-CoA transferase (BCoAT) or butyrate kinase (Buk) [8,9].

Fusobacterium nucleatum is a gram-negative anaerobic bacterium with species-specific reservoirs in the human oral cavity or gastrointestinal tract [10–12]. Recent accumulating evidence has shown that some *Fusobacterium*, especially *F. nucleatum*, are related to colorectal cancer progression [10,13]. In the cancer microenvironment, a decrease in *F. nucleatum* butyrate metabolism synergistically promotes inflammatory responses [13,14]. Whether there is a bidirectional effect between an *F. nucleatum*-mediated decrease of butyrate production and tumor development remains to be studied. Butyrate can be biologically reduced to butanol, during the solventogenic phase, via CoA transferase (CTF), aldehyde/alcohol dehydrogenase (AADH), and butanol dehydrogenase (BDH) (Figure 1) [15,16].

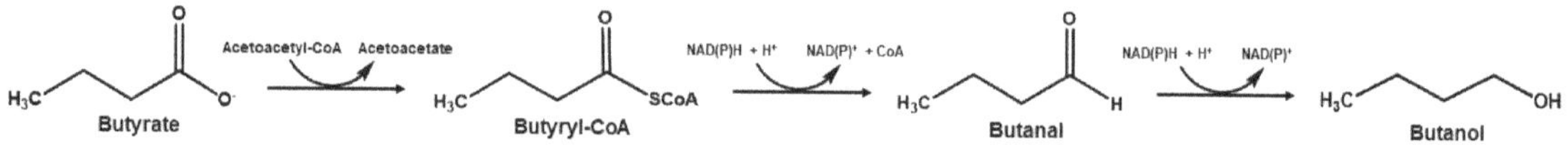

Figure 1. Butanol synthesis in *F. nucleatum*.

BDH plays a significant role in butanol production in many microorganisms by catalyzing butanal conversion to butanol at the expense of the NAD(P)H cofactor [17]. Homologs of BDH have been found in other anaerobic bacteria, including *Clostridium acetobutylicum* [18] and *Clostridium beijerinkii* [19], and they are involved in the n-butanol biosynthetic pathway. *F. nucleatum* possesses a putative *yqdH* gene that encodes the BDH protein and plays an essential role in butanol biosynthesis and butyrate metabolism [20]. Consequently, *F. nucleatum* YqdH (named FnYqdH) is likely to be a potential drug target for developing new types of antibiotics. However, the structure and function of FnYqdH have not been studied yet, and the molecular mechanism of butanol synthesis in anaerobic bacteria remains to be explored in detail.

In this study, we found that FnYqdH exhibited dual cofactor specificity for nicotinamide adenine dinucleotide (NADH) and nicotinamide adenine dinucleotide phosphate (NADPH) and has butanol dehydrogenase activity for various aldehydes. In addition, four metal ions, namely Co^{2+}, Ni^{2+}, Zn^{2+}, and Fe^{3+}, significantly enhanced the activity of metal-free FnYqdH. Moreover, FnYqdH showed broad substrate specificity toward aldehydes. To gain structural insights into the catalytic mechanism, along with substrate and cofactor specificities of FnYqdH, we determined the crystal structures of YqdH from *F. nucleatum* in its apo and complexed forms (with NADH and Co^{2+}). We found several residues involved in the coenzyme- and metal ion-binding of FnYqdH. Our study on FnYqdH could offer valuable clues for a better understanding of the molecular functions of BDH, which is a potential antibacterial target for *F. nucleatum*-related disease.

2. Results

2.1. Biochemical Analysis of FnYqdH

BDHs are classified as BDHA and BDHB based on their use of NADH and NADPH as cofactors, respectively [21]. In the GenBank, FnYqdH (accession number: AAL95608.1) has been annotated as NADH-dependent BDHA, indicating that it catalyzes butanal conversion to butanol using the NADH cofactor. BDHs essentially require a metal ion for substrate recognition during a dehydrogenase reaction, and they play a pivotal role in enzymatic activity [22–24]. To investigate the cofactor specificity and examine the effect of a metal ion on FnYqdH, the enzyme activities of native FnYqdH, EDTA-FnYqdH (metal-free; the metal of FnYqdH is removed by incubation with EDTA, a chelating agent), and that with the addition of Co^{2+} (native-Co^{2+} and EDTA-Co^{2+}, respectively) were measured using butanal as the substrate, with NADH or NADPH as cofactors, respectively. EDTA-FnYqdH had almost no BDH activity in the presence of NADH or NADPH, indicating that the metal ions are essential. Meanwhile, when adding Co^{2+} to EDTA-FnYqdH, the enzyme activity was recovered. When adding Co^{2+} to native FnYqdH, the enzyme activity was enhanced by 50–60% in the presence of NADH or NADPH, indicating that the functional metal ion sites were not fully occupied on purified FnYqdH. Notably, FnYqdH showed BDH activity with both NADH and NADPH (Figure 2a). In the presence of 4.5 μM FnYqdH, the consumption of NADH and NADPH was 417 μM and 357 μM, respectively, (Figure 2a and Figure S1), indicating the ability of FnYqdH to use both NADH and NADPH as cofactors. Furthermore, the apparent K_m and V_{max} values were measured with a purified protein. The results showed that NADH is preferred over NADPH in terms of FnYqdH activity (Figure S2). Subsequently, BDH activity was measured after incubating metal-free FnYqdH with metal ions (Mg^{2+}, Co^{2+}, Ni^{2+}, Zn^{2+}, and Fe^{3+}). Since Fe^{2+} is easily oxidized to Fe^{3+} at pH 6.0 [25], butanol dehydrogenase activity was monitored only in the presence of Fe^{3+} to avoid possible experimental errors when Fe^{2+} was used. The highest enzymatic activity of FnYqdH was observed in the presence of Co^{2+}; this activity was absent in the presence of Mg^{2+}. The overall BDH activity trend was as follows: $Ni^{2+} > Zn^{2+} > Fe^{3+}$, with relative activities of 93.7%, 85.2%, and 44.2%, respectively (Figure 2b).

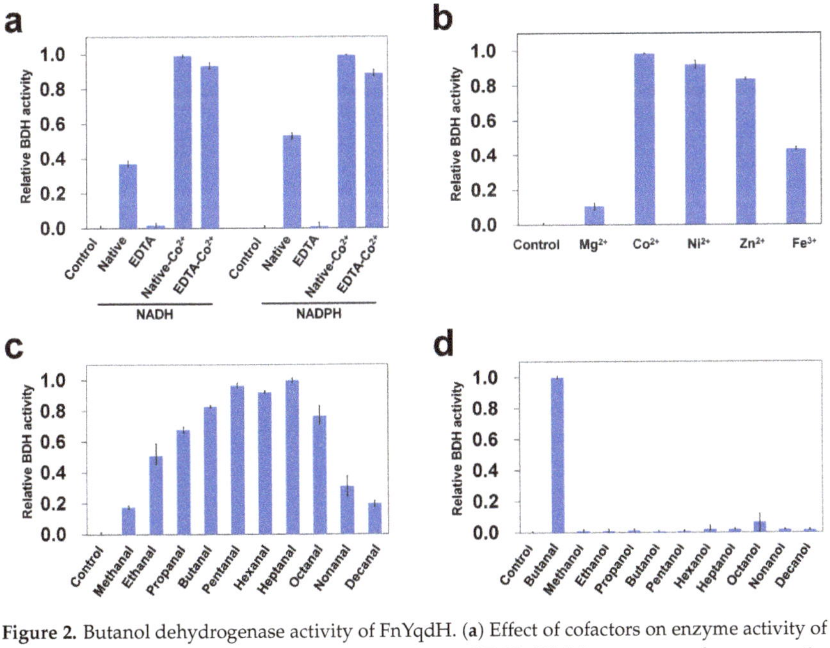

Figure 2. Butanol dehydrogenase activity of FnYqdH. (**a**) Effect of cofactors on enzyme activity of FnYqdH. (**b**) Effect of metal ions on enzyme activity of FnYqdH. Measurement of enzyme activity of FnYqdH toward the (**c**) aldehydes and (**d**) alcohols (the activity of butanal was defined as 100%). The graph represents the mean of three independent experiments, and error bars indicate the standard deviation.

To clarify the substrate specificity of FnYqdH, its activity was measured for ten different aldehydes (methanal, ethanal, propanal, butanal, pentanal, hexanal, heptanal, octanal, nonanal, and decanal). FnYqdH showed high activity for propanal (substrate degradation rate: 68%), butanal (83%), pentanal (96%), hexanal (92%), heptanal (100%), and octanal (77%) (Figure 2c); however, it showed very low enzyme activity for methanal (18%), ethanal (51%), nonanal (31%), and decanal (20%) (Figure 2c). Collectively, the results suggested that FnYqdH can convert aldehydes to alcohols, although the activity differed according to the length of the aliphatic chain—aldehydes with 3–8 carbon atoms in the aliphatic chain were preferred.

Protein BLAST analysis of FnYqdH revealed its high sequence similarity with several alcohol dehydrogenases (ADHs) from *Abyssisolibacter fermentans*(Accession code: WP_066501503.1, sequence identity: 57%), *Clostridium bovifaecis* (QGU94363.1, 55%), *Defluviitalea raffinosedens* (WSLF01000004, 54%), *Epulopiscium* sp. (NLK98329.1, 54%), and *Clostridiaceae bacterium* (NLY43117.1, 52%) (Figure S3). ADHs are ubiquitous enzymes that catalyze the oxidation of alcohols and reduction of aldehydes in the presence of $NAD(P)^+$ and NAD(P)H, respectively [26,27]. To determine the potential ADH activity of FnYqdH, the oxidase activity of FnYqdH was screened for ten different alcohols (methanol, ethanol, propanol, butanol, pentanol, hexanol, heptanol, octanol, nonanol, and decanol) in the presence of NAD^+. However, FnYqdH showed no significant enzymatic activity against any alcohol (Figure 2d), indicating that FnYqdH only has reductase activity for aldehydes, contrary to ubiquitous ADHs.

2.2. Overall Structure of FnYqdH

To better understand the molecular function of FnYqdH, we determined the crystal structure of FnYqdH using purified recombinant FnYqdH at 1.98 Å. However, metal ions and NAD(P)H were not observed in the electron-density map (named FnYqdH-apo). To identify the cofactor-binding form of FnYqdH, FnYqdH crystals were soaked for 1 min in a reservoir solution supplemented with NADH and Co^{2+}. After, we determined the crystal structure of FnYqdH complexed with NADH and Co^{2+} at 2.72 Å (named FnYqdH-NAD-Co^{2+}) (Table 1). The FnYqdH crystals belonged to the orthorhombic space group I222, containing one molecule in the asymmetric unit. The R_{work}/R_{free} of the final model of FnYqdH-apo and FnYqdH-NAD-Co^{2+} were 0.177/0.209 and 0.205/0.260, respectively.

Table 1. Crystallographic data and refinement statistics.

Data Collection	FnYqdH	FnYqdH-NAD-Co^{2+}
Beamline	Beamline 5C at PLS	Beamline 17B at SSRF
Resolution range (Å)	37.78–1.98 (2.05–1.98)	35.53–2.72 (2.82–2.72)
Space group	I222	I222
Total/unique reflections	39065 (3723)	14920 (1334)
a, b, c (Å)	64.77, 78.85, 215.22	64.426, 79.351, 212.932
α, β, γ (°)	90.00, 90.00, 90.00	90.00, 90.00, 90.00
CC1/2	0.975 (0.883)	0.929 (0.531)
CC*	0.994 (0.968)	0.981 (0.833)
Completeness (%)	99.58 (97.21)	98.64 (87.87)
Multiplicity	6.6 (6.0)	12.4 (10.3)
Average $I/\sigma(I)$	15.26 (1.8)	11.88 (1.6)
Refinement		
R_{work}/R_{free} (%)	17.7/20.9	20.5/26.0
No. of non-hydrogen atoms	3342	3121
Protein residues	384	384
B-factor (Å^2)	29.01	59.10
Protein	28.199	58.71
NADH		94.57
Co^{2+}		117.00
Water	36.97	50.00
R.m.s.d. from ideal		
RMS (Bond)	0.008	0.009
RMS (Angles)	0.81	1.61
Ramachandran plot (%)		
Favored regions	98.69	93.72
Allowed regions	1.31	6.02
Disallowed regions	0.00	0.26
PDB code	6L1K	8I29

Values in parentheses are for the outermost shell.

The monomer FnYqdH structure consisted of an NAD-binding domain (NBD) and a catalytic domain (CD), connected by a 3_{10}–helix η3 (linker). Further, it has a deep cleft between NBD and CD (Figure 3a). The NBD (residues 1–180) had a pair of Rossmann-like α/β/α sandwich folds formed by six parallel β-sheets (β2 to β6 and β9) and surrounded by six α-helices (α1 to α6) (Figure 3b). The two anti-parallel strands, β7 and β8, formed a long β-hairpin and were flanked by two 3_{10}-helices, η1 and η2. The CD (residues 190–385) contained nine α-helices and two 3_{10}-helices, η4 and η5, thereby representing typical dehydrogenase synthase-like α domain topology (Figure 3b). B-factor analysis of FnYqdH revealed that the NBD region (B-factor: 35.00 $Å^2$) was more flexible than the CD region (22.40 $Å^2$) while mobile loops gathered on the surface (Figure S4). Compared to B-factors of the main chain (29.01 $Å^2$), B-factor analysis further showed that most regions form a rigid structure, excluding the L4 (Lys32–Thr34, B-factor: 54.50 $Å^2$), α2 (Glu45–Leu50, 51.70 $Å^2$), L5 (Glu41–Gly44, 47.10 $Å^2$), and L15 (Glu160–Lys163, 63.40 $Å^2$). The conserved surface of FnYqdH exhibited the substrate-binding cleft as highly conserved, while other regions had no or relatively low amino acid conservation (Figure 3c).

In the crystal structure, two molecules of FnYqdH formed a dimer in symmetric crystal packing (Figure 3d), which was consistent with the dimeric state of FnYqdH as evidenced by size-exclusion chromatography (Figure 3e). The dimeric formation of FnYqdH was stabilized via an anti-parallel β-sheet formed between the N-terminal β1 of one molecule and β2′ strands of the second (Figure 3d). Dimeric interface analysis revealed the hydrogen bonds between three residue pairs (Asn3-Lys16*, Phe4-Phe14*, Tyr6-Ile12*, with * denoting the partner molecule) (Figure 3d). The buried surface area of the dimeric interface of FnYqdH was 1613.40 $Å^2$.

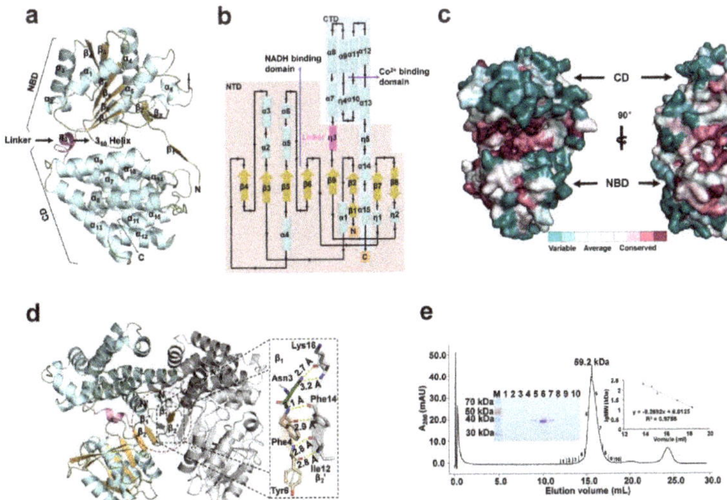

Figure 3. Crystal structure of FnYqdH. (**a**) Cartoon representation of monomer structure of FnYqdH consisting of the NBD, CD, and linker. (**b**) Topology diagram of FnYqdH. The α-helices (cyan) and β-strands (yellow) are indicated by rectangles and arrows, respectively. The smaller numbers indicate the beginning and ending residue number of each secondary structural element. The NTD is shadowed by light orange boxes. The CTD of FnYqdH is marked by a light cyan box. The linker is marked by a pink box. (**c**) Surface conservation of FnYqdH. The substrate-binding cleft between NBD and CD shows high conservation, whereas other regions are not conserved. (**d**) Dimer formation of FnYqdH. The hydrogen bonds between β1 and β2′ strands stabilize the dimeric interface of FnYqdH. (**e**) Profile of size-exclusion chromatography of FnYqdH revealed the dimer state in solution.

2.3. Substrate- and Cofactor-Binding Site of FnYqdH

As previously mentioned, both NADH and metal ions are required for the enzyme activity of FnYqdH. However, the detailed mechanism of the interaction of FnYqdH with cofactors remains unknown. In the crystal structure of FnYqdH-NAD-Co^{2+}, an electron-density map corresponding to NADH and metal ions was clearly observed (Figure S5). NADH was located on the NBD, which was coordinated by the O atom of Pro75 (distance: 3.50 Å), OG atom of Thr143 (3.20 Å), OG atom of Ser152 (3.10 Å), and O atom of Val184 (2.70 Å). In addition, a water molecule acting as a bridge between NADH and two residues (Glue41 and Ser104) was coordinated by the OE1 and OE2 atoms of Glu41 (3.30 Å and 3.50 Å) and the OG atom of Ser104 (3.20 Å) (Figure 4a). The superposition of NADH-binding residues of FnYqdH-apo with those of FnYqdH-NAD-Co^{2+} showed that Pro75, Ser152, and Val184 do not undergo conformational changes, although Glu41, Ser104, and Thr143 were positioned differently in both FnYqdH-apo and FnYqdH-NAD-Co^{2+} (Figure 4a). Sequence alignment of FnYqdH with other BDHs from four different species, namely *C. beijerinkii*, *C. acetobutylicum*, *Bacillus megaterium*, and *Zymomonas mobilis*, showed Pro75, Ser104, and Thr143 to be highly conserved in all the selected species (Figure S6). However, Glu41, Ser152, and Val184 of FnYqdH were not conserved, suggesting that in FnYqdH, three residues—namely Pro75, Ser104, and Thr143—are significant for the NADH-binding of BDH. In addition, we found 12 residues (Gly40, Gly41, Ser42, Gly99, Ser100, Asp103, Thr139, Asn148, Asn150, Lys161, Val180, and Thr183) to be crucial for NADPH binding in the crystal structure of TmBDH (homolog protein of FnYqdH, described below). To identify the key residue involved in NAD(P)H binding, we produced nine FnYqdH mutant proteins (E41A, G44A, L46A, S104A, D107A, T143A, S152A, K165A, and T187A) in *Escherichia coli*, using the same procedure used for wild type FnYqdH (FnYqdH-WT), and performed enzyme activity studies. The results showed the enzyme activities of the L46A, S104A, D107A, T143A, S152A, and T187A mutants to be completely abolished. In contrast, the enzyme activities of the E41A, G44A, and K165A mutants were retained as compared with other FnYqdH mutants (Figure 4b), indicating that Leu46, Ser104, Asp107, Thr143, Ser152, and Thr187 are critical residues for NADH recognition.

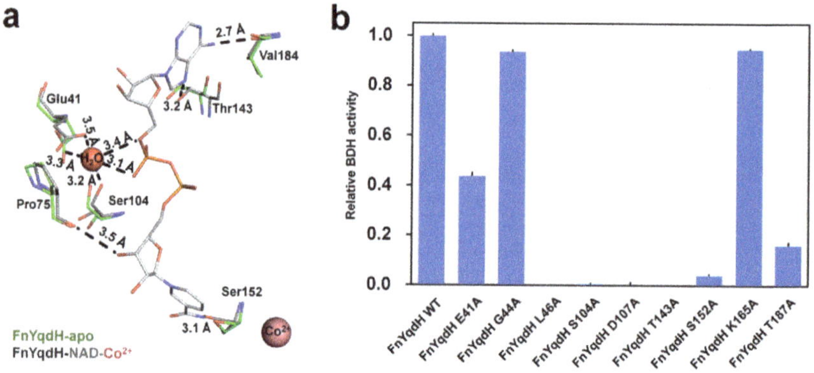

Figure 4. NADH-binding site in FnYqdH. (**a**) Superposition of the NADH-binding site in FnYqdH (green) and FnYqdH-NAD-Co^{2+} (gray). (**b**) BDH activity of FnYqdH mutants for NADH-binding residues (E41A, G44A, L46A, S104A, D107A, T143A, S152A, K165A, and T187A). The activity of wild type FnYqdH (FnYqdH-WT) was defined as 100%.

To better understand the conformational change in proteins, we superposed the structures of FnYqdH-apo and FnYqdH-NAD-Co^{2+} and found both structures to be similar, with an r.m.s.d. of 0.375 Å (Figure S7). Notably, the crystal structures of FnYqdH-apo and FnYqdH-NAD-Co^{2+} revealed an open conformation, in which NBD and CD were formed at approximately 30° and 20°, respectively (Figure S8). Moreover, we found the β6-η1 loop

(L12) to be distanced from NADH in the crystal structure of FnYqdH-NAD-Co^{2+} compared to that of FnYqdH-apo (Figure S8). Next, we measured the distance between Ala103 on NBD and Gly196 on CD of FnYqdH-apo and FnYqdH-NAD-Co^{2+} and found the distance to be approximately 12.3 Å and 12.9 Å, respectively (Figure 5a,b). Upon NADH and metal binding to FnYqdH, NBD and CD revealed a narrower substrate-binding cleft compared to that of FnYqdH-apo.

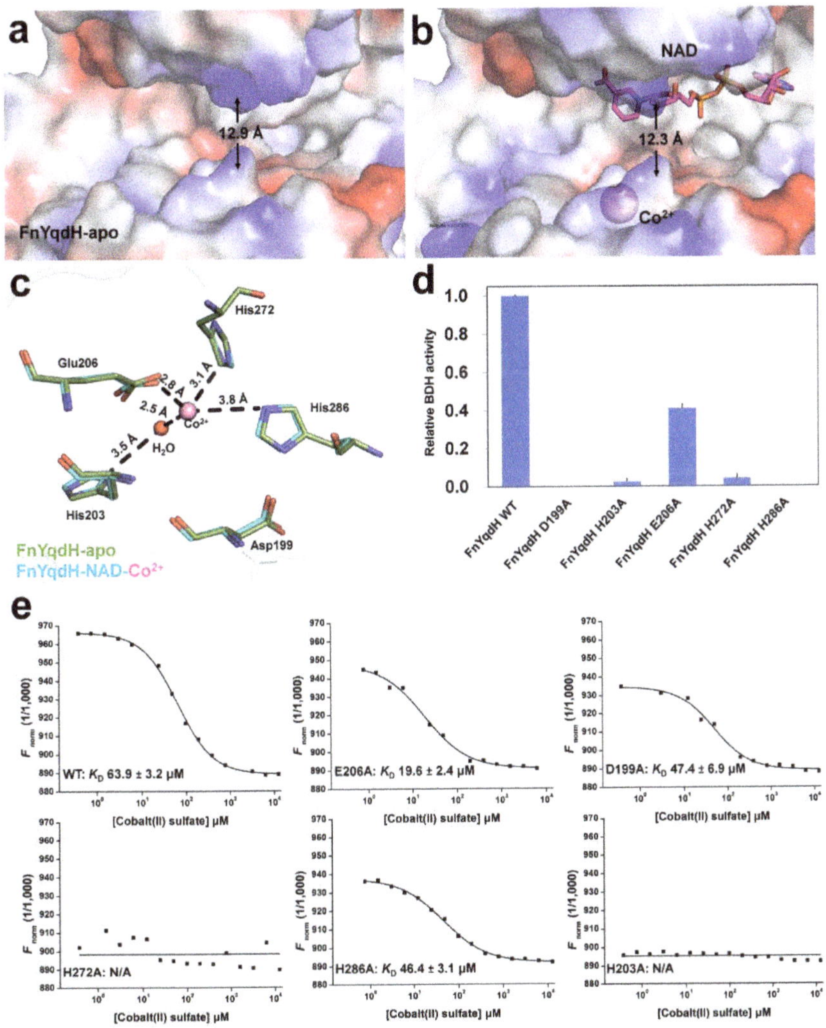

Figure 5. The metal-binding site of FnYqdH. The surface structure of substrate-binding cleft of (**a**) FnYqdH-apo and (**b**) FnYqdH-NAD-Co^{2+}. (**c**) Superposition of the Co^{2+}-binding site of FnYqdH (green) and FnYqdH-NAD-Co^{2+} (cyan). (**d**) BDH activity of FnYqdH with mutated metal-binding residues (FnYqdH-D199A, FnYqdH-H203A, FnYqdH-E206A, FnYqdH-H272A, and FnYqdH-H286A). The activity of FnYqdH-WT was defined as 100%. (**e**) MST assays of the Co^{2+}-binding affinity of FnYqdH-WT and its mutants (FnYqdH-D199A, FnYqdH-H203A, FnYqdH-E206A, FnYqdH-H272A, and FnYqdH-H286A). No Co^{2+}-binding affinity of FnYqdH-H203A and FnYqdH-H272A was detected. K_D values 63.9 μM, 19.6 μM, 47.4 μM, and 46.4 μM for Co^{2+}, FnYqdH-WT, FnYqdH-E206A, FnYqdH-D199A, and FnYqdH-H286A, respectively.

The electron-density map corresponding to Co^{2+} was observed around the L22 loop and α7-helix. Co^{2+} was coordinated by the OE1 atom of Glu206 (2.80 Å), ND1 atom of His272 (3.10 Å), NE2 atom of His286 (3.80 Å), and a water molecule (2.50 Å), thereby establishing tetrahedral coordination (Figure 5c). In addition, a water molecule acted as a bridge between Co^{2+} and the ND1 atoms of His203 (3.50 Å) (Figure 5c). B-factor analysis revealed the temperature factors of Co^{2+} (117.00 Å^2) to be higher than the average temperature factor of all atoms in the protein (59.10 Å^2), thereby indicating that Co^{2+} loosely interacts with FnYqdH. To clarify, the catalytic metal-binding residues of FnYqdH and the Co^{2+}-binding residues were substituted with alanine residues, and enzyme activity was measured. BDH activity of D199A, H203A, H272A, and H286A mutants was completely abolished, whereas the enzyme activity of the E206A mutant was retained partially as compared with other FnYqdH mutants (Figure 5d).

The Co^{2+}-binding affinity of FnYqdH-WT and the mutant proteins was measured using a microscale thermophoresis (MST) assay. The results showed that the K_D of WT, D199A, E206A, and H286A mutants was 63.9 μM, 47.4 μM, 19.6 μM, and 46.4 μM, respectively (Figure 5e). In contrast, no Co^{2+} binding was detected in H203A and H272A mutants (Figure 5e). Based on the biochemical and structural analyses of metal-binding residues, we concluded that His203 and His272 of FnYqdH are key residues for metal ion binding. Although Asp199 and His286 are not involved in metal ion binding, and these two residues play a pivotal role in FnYqdH activity.

2.4. Comparison of FnYqdH with Structural Homolog Proteins

To understand the molecular function of FnYqdH, homologous structures of FnYqdH were searched. The results showed that FnYqdH has structural homology with TmBDH (Protein Data Bank [PDB] code: 1VLJ; sequence identity = 40%; Z-score = 46.4), which is complexed with NADPH and Fe^{3+} (unpublished data). The domain organization and overall fold of TmBDH exhibited a high similarity to those of FnYqdH, but the superposition of FnYqdH-NAD-Co^{2+} and TmBDH structures illustrated a conformational difference (Figure 6a). In addition, the surface structure of FnYqdH-NAD-Co^{2+} revealed open conformation between the NBD and CD domains (approximately 20°), whereas TmBDH showed closed conformation between the NBD and CD domains (Figure 6b). In TmBDH, NADPH was stabilized via hydrogen bonds with 12 residues (Gly40, Gly41, Ser42, Gly99, Ser100, Asp103, Thr139, Asn148, Asn150, Lys161, Val180, and Thr183) (Figure 6c). Sequence alignment showed Gly44, Ser104, Asp107, Thr143, Lys165, Val184, and Thr187 of FnYqdH to be highly conserved in both FnYqdH and TmBDH, whereas Gly41, Leu46, Ala103, Ser152, and Ser154 of FnYqdH were not conserved in TmBDH (Figure 6d). Structure comparison of FnYqdH and TmBDH showed that some conserved amino acid residues (Gly44, Ser104, Asp107, and Lys165 of FnYqdH) have different conformations (Figure 6c).

In the crystal of TmBDH, Fe^{3+} was coordinated with the OD1 atom of Asp195 (2.80 Å), NE2 atom of His199 (2.20 Å), NE2 atom of His268 (2.30 Å), and NE2 atom of His282 (2.20 Å), forming tetrahedral coordination (Figure 6e). This metal ion position of TmBDH differs from that of Co^{2+} in FnYqdH. In addition, Asp199, His203, Glu206, His272, and His286 of FnYqdH had different conformations when compared with TmBDH. In particular, Asp195 and His199 of TmBDH directly interacted with Fe^{3+}, whereas His203 of FnYqdH relied on a water molecule for Co^{2+}, and Glu206 of FnYqdH interacted with Co^{2+} instead of Asp199 (Figures 5c and 6e). To understand the effect of different metal ions on the enzyme activity of metal-free TmBDH (the metal of TmBDH is removed by incubation with EDTA, a chelating agent), the enzyme activity was measured using a spectrophotometric assay for monitoring NADH consumption for five different metal ions (Mg^{2+}, Co^{2+}, Ni^{2+}, Zn^{2+}, and Fe^{3+}). The highest enzyme activity of TmBDH was observed in the presence of Zn^{2+}. The overall BDH activity trend was as follows: $Co^{2+} > Ni^{2+} > Fe^{3+}$, with relative activities of 91.2%, 80.1%, and 23.8%, respectively. Accordingly, the enzyme activity of TmBDH was enhanced by the addition of Zn^{2+}, Co^{2+}, and Ni^{2+}; however, Fe^{3+} did not significantly increase the activity, which was contrary to that in FnYqdH (Figure 6f). Moreover, we found that the NADH consumptions of TmBDH were 473 μM, 432 μM, and 381 μM in the presence of Zn^{2+}, Co^{2+}, and Ni^{2+}, respectively (Figure 6f and Figure S1a). Accordingly, based on our biochemical study, we determined that the crystal structure of Fe^{3+}-bound TmBDH is not the most active conformation in the metal-binding site.

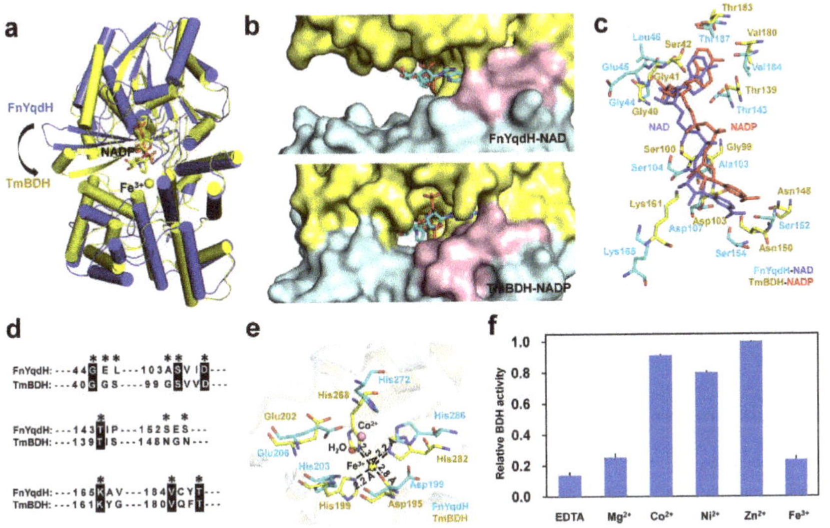

Figure 6. Structural and functional comparison of FnYqdH with TmBDH. (**a**) Superposition of monomeric structures of FnYqdH (blue) and TmBDH (yellow). (**b**) Comparison of the surface structures of FnYqdH-NAD-Co^{2+} and TmBDH. (**c**) Superposition of NAD(P)H-binding sites of FnYqdH-NAD-Co^{2+} (cyan) and TmBDH (yellow). NADH and NADPH are colored dark blue and red, respectively. (**d**) Partial sequence alignment of FnYqdH and TmBDH. The NADPH-binding site of TmBDH is indicated by an *. A black background highlights conserved residues. (**e**) Superposition of the metal ion-binding sites of FnYqdH-NAD-Co^{2+} (cyan) and TmBDH (yellow). Co^{2+}, Fe^{3+}, and water molecules are colored pink, red, and yellow, respectively. The dark dashed lines indicate interactions between Fe^{3+} and amino acids in TmBDH. (**f**) Effect of metal ions on the enzymatic activity of TmBDH.

3. Discussion

Butanol has important industrial applications as a fuel additive and a raw chemical for the plastics industry [28,29] and as a grade extractant in the food and flavor industry [30]. BDH plays a key role in butanol biosynthesis by converting butanal to butanol, using NAD(P)H as a cofactor [31]. However, current research on BDH largely focuses on non-pathogenic bacteria, such as *C. acetobutylicum* BDH and *C. beijerinkii* BDH, while several reports focus on BDH within pathogenic bacteria. To investigate whether the enteropathogens *Clostridioides difficile* [32] and *Clostridium perfringens* [33] use the same butyrate metabolism mechanism as *F. nucleatum*, we examined their butyrate metabolism pathways using KEGG [34]. Through KEGG analysis, we found that in *C. difficile*, butyrate can be reduced to butanol during butanoate metabolism in two ways:first, via the 3-oxoacid CoA-transferase, aldehyde/alcohol dehydrogenase, and alcohol dehydrogenase pathway, and, second, via the butyrate kinase, phosphotransbutyrylase, aldehyde/alcohol dehydrogenase, and alcohol dehydrogenase pathway (Figure S9) [34]. Moreover, we found that butyrate can be reduced to butanol during butanoate metabolism via the butyrate kinase, phosphate butyryltransferase, aldehyde/alcohol dehydrogenase, and alcohol dehydrogenase pathway in *C. perfringens* (Figure S10) [34]. Therefore, *C. difficile* has the same butyrate metabolic pathway as *F. nucleatum*, whereas the conversion of butyrate to butyrate-CoA in *C. perfringens* involves a different enzyme.

In this study, we first characterized BDH from *F. nucleatum*, which is an emerging human bacterial pathogen, and determined the crystal structure of FnYqdH in its apo and complexed forms (with NADH and Co^{2+}). We also characterized TmBDH, whose crystal structure had been deposited in the PDB by another research group (unpublished). The BDH family is classified into BDHA and BDHB based on the cofactors NADH and NADPH, respectively. Interestingly, FnYqdH exhibited BDH activity using both NADH and NADPH. Moreover, we found that the use of both NADH and NADPH as cofactors is not distinctive to FnYqdH. Yao et. al. found that the bifunctional alcohol–aldehyde dehydrogenase (AdhE) of *Thermoanaerobacter mathranii* showed a small amount of NADPH activity in addition to NADH activity [35]. Pei et. al. found that the AdhE of *Thermoanaerobacter ethanolicus* showed NADH aldehyde dehydrogenase activity and small amounts of NADPH alcohol dehydrogenase activity [36]. Biswas et. al. found that the Asp-494-Gly (D494G) mutation in AdhE of *Clostridium thermocellum* causes the enzyme to use both NADH and NADPH as cofactors, while WT prefers to use only NADH as a cofactor [37,38]. Further, the AdhE cofactor specificity is thought to be determined by the presence or absence of electrostatic repulsion and steric hindrance in the 2′-phosphate binding pocket of NADPH [38]. In the crystal structure of FnYqdH-NAD-Co^{2+}, the 2′-phosphate of NADPH (aligned with TmBDH) is not sterically hindered in the binding pocket, which implies that NADPH can enter the binding pocket (Figure S11). Thus, FnYqdH exhibits bifunctional cofactor specificity.

FnYqdH has broad substrate specificity, showing the highest activity with heptanal and lowest with methanal. FnYqdH can convert aldehyde to alcohol by utilizing NAD(P)H with broad substrate specificity; however, it has no enzyme activity toward alcohols. Accordingly, the enzyme activity of FnYqdH differed from that of typical ADH, which catalyzes the oxidation of alcohols and reduction of aldehydes [26]. These findings indicate that the reductase activity of FnYqdH is useful for producing biorenewable fuels and chemicals, such as butanol, pentanol, hexanol, and heptanol. The in vitro biochemical study herein revealed that FnYqdH shows BDH activity in the presence of Co^{2+}, Ni^{2+}, Zn^{2+}, and Fe^{3+}, with Co^{2+} being associated with the highest activity. Interestingly, even though Fe^{3+} is bound to the crystal structure of TmBDH, TmBDH showed low activity with Fe^{3+} and high activity in the presence of Zn^{2+}. Since our experiments on optimal metal ions for BDH activity were performed in vitro, further in vivo experiments on the preferred metals are warranted. Combining these biochemical experiments of FnYqdH and TmBDH suggests two useful experimental directions for the industrial application of BDH. First, since BDH activity changes in vitro depending on the type of metal, the target BDH for industrial application can increase productivity by screening and selecting metals with optimal activity. Second, the substrate specificity of BDHs is not limited to butanal. BDHs can be applied to convert various aldehydes into alcohols, highlighting the potential for the industrial production of butanol and other alcohols.

Crystal structures of FnYqdH-apo and FnYqdH-NAD-Co^{2+} showed an open conformation between NBD and CD, with a broad substrate-binding cleft of more than 12 Å present between the two domains. The distance between the two domains was reduced by approximately 1 Å upon binding NADH and Co^{2+}, indicating that the cofactor did not significantly affect the conformation between the two domains of FnYqdH. For the BDH activity of FnYqdH, it is essential that NAD and metal ions involved in the activity are in close proximity. Indeed, for the two cofactors to be close, the two domains must be close. Accordingly, in the crystal structures of FnYqdH determined in this experiment, the open conformation corresponded to the pre-substrate binding state and not the close conformation of the active state. Meanwhile, FnYqdH preferred aldehydes with an aliphatic chain length of approximately 3–8, thereby indicating that the aliphatic chain of aldehyde can interact with FnYqdH. As a result, we expect that the open conformation between the NBD and CD domains of FnYqdH will change to a closed conformation when FnYqdH binds to a substrate or undergoes a BDH reaction. In contrast, the crystal structure of TmBDH showed a closed conformation, maintaining a close distance to the nucleotide and metal compared to FnYqdH; hence, it could be determined whether it was very close to the active state. To understand the substrate recognition and catalytic mechanisms, the crystal structure of FnYqdH complexed with cofactors and substrate is required.

Moreover, based on structural analyses, we identified key residues involved in NADH and metal binding through a comprehensive study, including mutagenesis, biochemical assay for BDH activity, and MST assays. The results of these mutagenesis and biochemical assays provide a deeper understanding of the molecular properties of the BDH family. However, the substrate-binding site remains unknown. The ADH of *C. acetobutylicum* (CaADH) consists of an N- and a C-terminal domain, which form two clefts. NADH occupies one cleft, and the other forms a substrate-binding chamber [39]. Structural comparison of FnYqdH-NAD-Co^{2+} and a model structure of CaADH, *Geobacillus thermoglucosidasius* ADH (PDB code: 3ZDR, r.m.s.d.: 2.503 Å), showed that FnYqdH has a substrate-binding chamber located in a similar position (Figure S12). In addition, Rellos et. al. found that Ala161 is at the ADH of *Z. mobilis* (ZmADH) substrate-binding site, and its conversion to Val or Ile also allows butanol oxidization [40]. Sequence alignment of FnYqdH and ZmADH showed that Ala161 is not highly conserved, but in FnYqdH, it is converted to Val (Val167), which is a hydrophobic amino acid similar to Ala (Figure S12). Structural comparison of FnYqdH-NAD-Co^{2+} and ZmADH (PDB code: 3OX4, r.m.s.d.: 1.138 Å) showed that Val163 (FnYqdH) and Ala161 (ZmADH) are in similar positions (Figure S12). Sequence alignment of FnYqdH and other ADHs showed that most of the amino acids at this site are hydrophobic (Figures S5 and S8). Therefore, we determined that these results imply the direct involvement of Val167 of FnYqdH in substrate binding. Of course, further experiments should be carried out to verify this hypothesis.

In nature, various types of dehydrogenases are utilized in extended proton relay systems. Among them, the molecular mechanism of *Thermoanaerobacter brockii* ADH (TbADH) has been well established [41]. In TbADH, the binding of the alcohol substrate is considered to form a penta-coordinated zinc complex [41]. In the catalytic cycle, two supposed penta-coordinate intermediates are formed. Specifically, the tetra-coordinated zinc ion species is maintained when $NADP^{+}$ is bound to TbADH, and the first transient penta-coordinated zinc ion complex is formed by adding a water molecule [41]. The penta-coordinated zinc ion species is then reconverted to a tetra-coordinated zinc ion complex due to the dissociation of ligated Glu60 residue. When the alcohol substrate is bound, a penta-coordinated zinc ion complex is formed, which corresponds to the second transient species. When the water molecule and product dissociate, the original tetra-coordinated zinc ion species are regenerated by re-ligating Glu60, and the catalytic cycle is complete, returning the enzyme to its resting state [41]. By active site analysis of FnYqdH, we found Glu206 to be located nearby metal ions and TbADH; this residue was highly conserved in other BDHs (Figure S8). Site-directed mutagenesis and MST assay revealed that the enzyme activity of FnYqdH-E206A was reduced, although metal-binding affinity persisted. The results suggested that Glu206 of BDH is involved in the catalytic process and interacts weakly with metal ions.

Based on previous reports and our results, we proposed an alternative model for a putative proton relay of FnYqdH. When NADH and a water molecule are bound to FnYqdH, a transient tetra-coordinated Co^{2+} complex is formed. The binding of the incoming aldehyde substrate forms a penta-coordinated Co^{2+} complex. The latter is then transformed into a tetra-coordinated Co^{2+} complex by dissociating the ligated Glu206 residue. The catalytic cycle is envisaged to restart by binding the water molecule and NADH (Figure 7). Based on the results above, we considered that further experimental investigations would be required to ensure whether the systems would indeed apply to other BDHs.

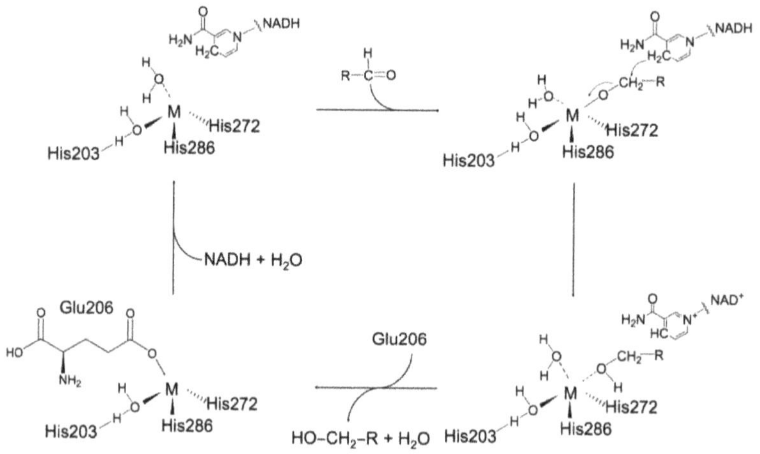

Figure 7. Proposed mechanism of BDH activity of FnYqdH. A tentative reaction mechanism involves extended proton relay systems.

4. Materials and Methods

4.1. Cloning, Expression, and Purification

Genomic DNA from *F. nucleatum* was used as a template to amplify the yqdH gene (GenBank: AAL95608.1) by a conventional polymerase chain reaction (PCR). The PCR fragment was digested at the NcoI and XhoI restriction sites and cloned into the pPRO-EX-HTA vector. Expression and purification of FnYqdH were performed as previously described [42]. Site-directed mutagenesis was performed by two subsequent PCR reactions using the primers shown in Table S1. Mutant proteins were expressed and purified using the same procedure as the FnYqdH-WT.

4.2. Enzyme Activity Assays

Enzyme activity of FnYqdH was determined using butanal as a substrate to measure NAD(P)H consumption. The reaction mixture contained 10 mM Tris-HCl, pH 6.0, 0.5 mM NADH or NADPH, 50 mM butanal, 2 mM $CoSO_4$, and an enzyme in a reaction volume of 100 μL. The reaction mixture was incubated for 10 min at 35 °C and then quickly placed in an ice bath for 5 min, after which the reaction was terminated. Enzyme activities were then assayed by monitoring NAD(P)H consumption using a microplate spectrophotometer at 340 nm. All experimental data are averages of at least triplicates.

4.3. Standard Curve of NADH and NADPH

The NAD(P)H concentration curve was measured using a microplate spectrophotometer at 340 nm. Serial dilutions of NAD(P)H ranging from 0–700 μM were made using 10 mM Tris-HCl, pH 6.0, 2 mM $CoSO_4$, and 50 mM butanal. All experimental data are averages of at least triplicates.

4.4. Metal Substitution

Purified FnYqdH and TmBDH proteins were each diluted two-fold with chelating buffer containing 20 mM Tris-HCl, pH 8.0, 150 mM NaCl, and 10 mM EDTA and incubated at 25 °C for 1 h. The protein solution was further dialyzed using a buffer containing 20 mM Tris-HCl, pH 8.0, and 150 mM NaCl at 25 °C overnight. Further, metal-free FnYqdH and TmBDH proteins were incubated with 2 mM metal ions ($ZnCl_2$, $MgCl_2$, $CoSO_4$, $FeCl_3$, and $NiCl_2$).

4.5. Substrate Spectrum

Substrate specificity of FnYqdH was determined. The activity of FnYqdH in reducing methanal, ethanal, propanal, butanal, pentanal, hexanal, heptanal, octanal, nonanal, and decanal was detected by the standard method. The reaction system (100 μL) usually consisted of the corresponding substrate, as mentioned above, in 10 mM Tris-HCl, pH 6.0, 0.5 mM NADH, 2 mM $CoSO_4$, and an appropriate amount of enzyme.

The activity of FnYqdH in the oxidation of methanol, ethanol, propanol, butanol, pentanol, hexanol, heptanol, octanol, nonanol, and decanol using NAD^+ as a coenzyme was used to measure NADH production. The reaction system (100 μL) usually consisted of the corresponding substrate, as mentioned above, in 10 mM Tris-HCl, pH 6.0, 0.5 mM NAD^+, 2 mM $CoSO_4$, and an appropriate amount of enzyme. It was then assayed by monitoring NADH production using a microplate spectrophotometer at 340 nm. All experimental data are averages of at least triplicates.

The double reciprocal plot was applied to obtain the kinetic parameters of FnYqdH in the presence of NADH (0.1–0.5 mM) or NADPH (0.1–1 mM).

4.6. Crystallization and Data Collection

Crystallization and preliminary X-ray diffraction analysis of FnYqdH-apo have been published previously [42]. High-quality FnYqdH crystals were incubated for 1 min in a solution consisting of 25% (*v*/*v*) glycerol. X-ray diffraction data were collected with a Beamline 5C from the Pohang Light Source (PLS, Pohang, Republic of Korea) at −173 °C. Diffraction images were recorded with an ADSC Q315 CCD detector and processed with the HKL2000 program. For FnYqdH complexed with NADH and Co^{2+}, a crystal of FnYqdH was soaked in a reservoir solution consisting of 25% (*v*/*v*) glycerol, 10 mM NADH, and 2 mM Co^{2+} and incubated for 1 min. X-ray diffraction data were collected at beamline BL17B of the Shanghai Synchrotron Radiation Facility (SSRF, Shanghai, China) at −173 °C. Diffraction images were recorded with a Rayonix MX300 detector (Rayonix, LLC, Evanston, IL, USA) and processed with the HKL2000 program [43,44].

4.7. Structure Determination and Refinement

The crystal structure of FnYqdH was determined by the molecular replacement (MR) method using the Phenix program [45,46]. The crystal structure of TmBDH (PDB code: 1LVJ) was used as a search template model. Model building and refinement were conducted with the Crystallographic Object-Oriented Toolkit (COOT) [47] and Phenix [45]. The structure of FnYqdH-NAD-Co^{2+} was determined by an MR method, using the native structure of FnYqdH as the search model. Ligands and water molecules were added by manually inspecting a 2Fo-Fc electron-density map as a guide in COOT [47]. All figure structures were generated using the program PyMOL [48]. The geometry of the final models was validated with MolProbity [49]; homologous structures of FnYqdH were searched using the DaliLite server [50].

4.8. Size-Exclusion Chromatography

To determine the molecular weight and the oligomeric state of FnYqdH, purified FnYqdH (500 μL of 1 mg/mL) was loaded onto a Superdex 200 10/300 GL column (GE Healthcare, Chicago, IL, USA), and the protein was eluted using a buffer containing 20 mM Tris-HCl, pH 8.0, 150 mM NaCl, and 2 mM β-mercaptoethanol (BME), at a flow rate of 0.5 mL/min.

4.9. Labeling and MST Measurements

The binding constant of FnYqdH, with respect to the substrate, was measured using MST. Further, 400 nM of purified FnYqdH and mutant proteins and 100 nM dye were diluted in assay buffer (20 mM HEPES, pH 8.0, and 0.5% polysorbate 20). The protein and the dye were mixed and incubated at 25 °C for 30 min. After incubation, the supernatant was collected by centrifugation at 13,000 rpm for 15 min at 4 °C. Increasing concentrations

(380 nM to 12.5 mM) of $CoSO_4$ in the assay buffer were mixed with constant amounts of labeled FnYqdH or mutant proteins. Samples were loaded into standard capillaries (Monolith NT.115), and all measurements were performed at 40% LED power and medium MST power, in duplicates. Data evaluation and K_D value determination were performed using NT analysis software (Nano Temper Technologies, Munich, Germany) [51].

5. Conclusions

We comprehensively analyzed the molecular function of FnYqdH using biochemical studies, structure analyses, and mutagenesis. Based on our and previously reported results, we proposed a putative proton relay system for the BDH activity of FnYqdH. Our results expand knowledge of the BDH family, thereby providing insight into the future application of BDH in industry.

Supplementary Materials: The following supporting information can be downloaded at: https://www.mdpi.com/article/10.3390/ijms24032994/s1.

Author Contributions: Conceptualization, K.H.N. and Y.X.; methodology, X.B., J.L., T.B., C.Q., K.H.N. and Y.X.; formal analysis, X.B., J.L., I.J., L.Z., K.H.N. and Y.X.; supervision, C.Q., K.H.N. and Y.X.; validation, X.B., J.L., T.B., S.H., J.Z., L.W., Y.M., W.G., C.Q., K.H.N. and Y.X.; investigation, X.B., J.L., T.B., S.H., J.Z., X.J., C.Q., K.H.N. and Y.X.; writing—review and editing, M.L., H.P., C.Q., K.H.N. and Y.X.; resources, C.Q. and Y.X.; funding acquisition, K.H.N. and Y.X.; visualization, Y.X. All authors have read and agreed to the published version of the manuscript.

Funding: This work was supported by the natural science foundation of Liaoning province (grant no. 2019-MS-065 to Y.X.) and the National Research Foundation of Korea (Grant Nos. NRF-2017M3A9F6029736, NRF-2020M3H1A1075314 and NRF-2021R1I1A1A01050838 to K.H.N). This study was supported by ProGen (to K.H.N.).

Institutional Review Board Statement: Not applicable.

Informed Consent Statement: Not applicable.

Data Availability Statement: Coordinates and structure factors of FnYqdH and FnYqdH-NAD-Co^{2+} were deposited in RCSB PDB with accession codes 6L1K and 8I29, respectively.

Acknowledgments: The authors acknowledge the SSRF for their assistance with X-ray diffraction data collection.

Conflicts of Interest: The authors declare no conflict of interest.

References

1. Geirnaert, A.; Calatayud, M.; Grootaert, C.; Laukens, D.; Devriese, S.; Smagghe, G.; De Vos, M.; Boon, N.; Van de Wiele, T. Butyrate-producing bacteria supplemented in vitro to Crohn's disease patient microbiota increased butyrate production and enhanced intestinal epithelial barrier integrity. *Sci. Rep.* **2017**, *7*, 11450. [CrossRef] [PubMed]
2. Bordonaro, M.; Lazarova, D.L.; Sartorelli, A.C. Butyrate and Wnt signaling: A possible solution to the puzzle of dietary fiber and colon cancer risk? *Cell Cycle* **2008**, *7*, 1178–1183. [CrossRef] [PubMed]
3. Vital, M.; Howe, A.C.; Tiedje, J.M. Revealing the bacterial butyrate synthesis pathways by analyzing (meta)genomic data. *mBio* **2014**, *5*, e00889. [CrossRef]
4. Pryde, S.E.; Duncan, S.H.; Hold, G.L.; Stewart, C.S.; Flint, H.J. The microbiology of butyrate formation in the human colon. *FEMS Microbiol. Lett.* **2002**, *217*, 133–139. [CrossRef]
5. Anand, S.; Kaur, H.; Mande, S.S. Comparative In silico Analysis of Butyrate Production Pathways in Gut Commensals and Pathogens. *Front. Microbiol.* **2016**, *7*, 1945. [CrossRef]
6. Louis, P.; Flint, H.J. Diversity, metabolism and microbial ecology of butyrate-producing bacteria from the human large intestine. *FEMS Microbiol. Lett.* **2009**, *294*, 1–8. [CrossRef]
7. Chowdhury, N.P.; Mowafy, A.M.; Demmer, J.K.; Upadhyay, V.; Koelzer, S.; Jayamani, E.; Kahnt, J.; Hornung, M.; Demmer, U.; Ermler, U.; et al. Studies on the mechanism of electron bifurcation catalyzed by electron transferring flavoprotein (Etf) and butyryl-CoA dehydrogenase (Bcd) of *Acidaminococcus fermentans*. *J. Biol. Chem.* **2014**, *289*, 5145–5157. [CrossRef] [PubMed]
8. Louis, P.; Flint, H.J. Development of a semiquantitative degenerate real-time pcr-based assay for estimation of numbers of butyryl-coenzyme A (CoA) CoA transferase genes in complex bacterial samples. *Appl. Environ. Microbiol.* **2007**, *73*, 2009–2012. [CrossRef]

Vital, M.; Penton, C.R.; Wang, Q.; Young, V.B.; Antonopoulos, D.A.; Sogin, M.L.; Morrison, H.G.; Raffals, L.; Chang, E.B.; Huffnagle, G.B.; et al. A gene-targeted approach to investigate the intestinal butyrate-producing bacterial community. *Microbiome* **2013**, *1*, 8. [CrossRef] [PubMed]
). Brennan, C.A.; Garrett, W.S. *Fusobacterium nucleatum*—Symbiont, opportunist and oncobacterium. *Nat. Rev. Microbiol.* **2019**, *17*, 156–166. [CrossRef] [PubMed]
1. Kistler, J.O.; Booth, V.; Bradshaw, D.J.; Wade, W.G. Bacterial community development in experimental gingivitis. *PLoS ONE* **2013**, *8*, e71227. [CrossRef] [PubMed]
2. Yang, N.Y.; Zhang, Q.; Li, J.L.; Yang, S.H.; Shi, Q. Progression of periodontal inflammation in adolescents is associated with increased number of *Porphyromonas gingivalis*, *Prevotella intermedia*, *Tannerella forsythensis*, and *Fusobacterium nucleatum*. *Int. J. Paediatr. Dent.* **2014**, *24*, 226–233. [CrossRef]
3. Shang, F.M.; Liu, H.L. *Fusobacterium nucleatum* and colorectal cancer: A review. *World. J. Gastrointest. Oncol.* **2018**, *10*, 71–81. [CrossRef] [PubMed]
4. Wu, J.; Li, Q.; Fu, X. *Fusobacterium nucleatum* Contributes to the Carcinogenesis of Colorectal Cancer by Inducing Inflammation and Suppressing Host Immunity. *Transl. Oncol.* **2019**, *12*, 846–851. [CrossRef]
5. Al-Shorgani, N.K.N.; Shukor, H.; Abdeshahian, P.; Kalil, M.S.; Yusoff, W.M.W.; Hamid, A.A. Enhanced butanol production by optimization of medium parameters using *Clostridium acetobutylicum* YM1. *Saudi. J. Biol. Sci.* **2018**, *25*, 1308–1321. [CrossRef]
6. Wang, L.; Chauliac, D.; Moritz, B.E.; Zhang, G.; Ingram, L.O.; Shanmugam, K.T. Metabolic engineering of *Escherichia coli* for the production of butyric acid at high titer and productivity. *Biotechnol. Biofuels* **2019**, *12*, 62. [CrossRef]
7. Rajagopalan, G.; He, J.; Yang, K.-L. A Highly Efficient NADH-dependent Butanol Dehydrogenase from High-butanol-producing *Clostridium* sp. BOH3. *BioEnergy Res.* **2012**, *6*, 240–251. [CrossRef]
8. Lee, J.; Jang, Y.S.; Choi, S.J.; Im, J.A.; Song, H.; Cho, J.H.; Seung do, Y.; Papoutsakis, E.T.; Bennett, G.N.; Lee, S.Y. Metabolic engineering of *Clostridium acetobutylicum* ATCC 824 for isopropanol-butanol-ethanol fermentation. *Appl. Environ. Microbiol.* **2012**, *78*, 1416–1423. [CrossRef]
9. Hiu, S.F.; Zhu, C.X.; Yan, R.T.; Chen, J.S. Butanol-Ethanol Dehydrogenase and Butanol-Ethanol-Isopropanol Dehydrogenase: Different Alcohol Dehydrogenases in Two Strains of *Clostridium beijerinckii* (*Clostridium butylicum*). *Appl. Environ. Microbiol.* **1987**, *53*, 697–703. [CrossRef]
20. Kapatral, V.; Anderson, I.; Ivanova, N.; Reznik, G.; Los, T.; Lykidis, A.; Bhattacharyya, A.; Bartman, A.; Gardner, W.; Grechkin, G.; et al. Genome sequence and analysis of the oral bacterium *Fusobacterium nucleatum* strain ATCC 25586. *J. Bacteriol.* **2002**, *184*, 2005–2018. [CrossRef] [PubMed]
21. Walter, K.A.; Bennett, G.N.; Papoutsakis, E.T. Molecular characterization of two *Clostridium acetobutylicum* ATCC 824 butanol dehydrogenase isozyme genes. *J. Bacteriol.* **1992**, *174*, 7149–7158. [CrossRef] [PubMed]
22. Riordan, J.F. The role of metals in enzyme activity. *Ann. Clin. Lab. Sci.* **1977**, *7*, 119–129. [PubMed]
23. Sulzenbacher, G.; Alvarez, K.; Van Den Heuvel, R.H.; Versluis, C.; Spinelli, S.; Campanacci, V.; Valencia, C.; Cambillau, C.; Eklund, H.; Tegoni, M. Crystal structure of *E. coli* alcohol dehydrogenase YqhD: Evidence of a covalently modified NADP coenzyme. *J. Mol. Biol.* **2004**, *342*, 489–502. [CrossRef] [PubMed]
24. Plapp, B.V.; Savarimuthu, B.R.; Ferraro, D.J.; Rubach, J.K.; Brown, E.N.; Ramaswamy, S. Horse Liver Alcohol Dehydrogenase: Zinc Coordination and Catalysis. *Biochemistry* **2017**, *56*, 3632–3646. [CrossRef]
25. Morgan, B.; Lahav, O. The effect of pH on the kinetics of spontaneous Fe(II) oxidation by O_2 in aqueous solution–basic principles and a simple heuristic description. *Chemosphere* **2007**, *68*, 2080–2084. [CrossRef]
26. Larson, S.B.; Jones, J.A.; McPherson, A. The structure of an iron-containing alcohol dehydrogenase from a hyperthermophilic archaeon in two chemical states. *Acta Crystallogr. F. Struct. Biol. Commun.* **2019**, *75 Pt 4*, 217–226. [CrossRef]
27. Lu, H.; Yu, S.; Qin, F.; Ning, W.; Ma, X.; Tian, K.; Li, Z.; Zhou, K. A secretion-based dual fluorescence assay for high-throughput screening of alcohol dehydrogenases. *Biotechnol. Bioeng.* **2021**, *118*, 1624–1635. [CrossRef]
28. Green, E.M. Fermentative production of butanol—The industrial perspective. *Curr. Opin. Biotechnol.* **2011**, *22*, 337–343. [CrossRef]
29. Nair, P.; Meenakshi, H.N. Review on the synthesis, performance and trends of butanol: A cleaner fuel additive for gasoline. *Int. J. Ambient Energy* **2021**, *43*, 4207–4223. [CrossRef]
30. Lin, Y.; Wang, J.; Wang, X.; Sun, X. Optimization of butanol production from corn straw hydrolysate by *Clostridium acetobutylicum* using response surface method. *Chin. Sci. Bull.* **2011**, *56*, 1422–1428. [CrossRef]
31. Li, T.; Zhang, C.; Yang, K.L.; He, J. Unique genetic cassettes in a *Thermoanaerobacterium* contribute to simultaneous conversion of cellulose and monosugars into butanol. *Sci. Adv.* **2018**, *4*, e1701475. [CrossRef] [PubMed]
32. Gregory, A.L.; Pensinger, D.A.; Hryckowian, A.J. A short chain fatty acid-centric view of *Clostridioides difficile* pathogenesis. *PLoS Pathog.* **2021**, *17*, e1009959. [CrossRef] [PubMed]
33. Huang, C.Y.; Wang, M.C. *Clostridium perfringens* bacteremia associated with colorectal cancer in an elderly woman. *Turk. J. Gastroenterol.* **2020**, *31*, 960–961. [CrossRef] [PubMed]
34. Kanehisa, M.; Furumichi, M.; Sato, Y.; Kawashima, M.; Ishiguro-Watanabe, M. KEGG for taxonomy-based analysis of pathways and genomes. *Nucleic Acids Res.* **2023**, *51*, D587–D592. [CrossRef] [PubMed]
35. Yao, S.; Mikkelsen, M.J. Identification and overexpression of a bifunctional aldehyde/alcohol dehydrogenase responsible for ethanol production in *Thermoanaerobacter mathranii*. *J. Mol. Microbiol. Biotechnol.* **2010**, *19*, 123–133. [CrossRef]

36. Pei, J.; Zhou, Q.; Jiang, Y.; Le, Y.; Li, H.; Shao, W.; Wiegel, J. *Thermoanaerobacter* spp. control ethanol pathway via transcriptional regulation and versatility of key enzymes. *Metab. Eng.* **2010**, *12*, 420–428. [CrossRef]
37. Biswas, R.; Zheng, T.; Olson, D.G.; Lynd, L.R.; Guss, A.M. Elimination of hydrogenase active site assembly blocks H2 production and increases ethanol yield in *Clostridium thermocellum*. *Biotechnol. Biofuels* **2015**, *8*, 20. [CrossRef]
38. Zheng, T.; Olson, D.G.; Tian, L.; Bomble, Y.J.; Himmel, M.E.; Lo, J.; Hon, S.; Shaw, A.J.; van Dijken, J.P.; Lynd, L.R. Cofactor specificity of the bifunctional alcohol and aldehyde dehydrogenase (AdhE) in wild-type and mutant *Clostridium thermocellum* and *Thermoanaerobacterium saccharolyticum*. *J. Bacteriol.* **2015**, *197*, 2610–2619. [CrossRef]
39. Cho, C.; Hong, S.; Moon, H.G.; Jang, Y.S.; Kim, D.; Lee, S.Y. Engineering Clostridial Aldehyde/Alcohol Dehydrogenase for Selective Butanol Production. *mBio* **2019**, *10*, e02683–e02718. [CrossRef]
40. Rellos, P.; Ma, J.; Scopes, R.K. Alteration of Substrate Specificity of *Zymomonas mobilis* Alcohol Dehydrogenase-2 Usingin *Vitro*Random Mutagenesis. *Protein Expr. Purif.* **1997**, *9*, 83–90. [CrossRef]
41. Kleifeld, O.; Frenkel, A.; Martin, J.M.; Sagi, I. Active site electronic structure and dynamics during metalloenzyme catalysis. *Nat. Struct. Biol.* **2003**, *10*, 98–103. [CrossRef]
42. Bai, X.; Lan, J.; He, S.; Bu, T.; Zhang, J.; Wang, L.; Quan, C.; Nam, K.H.; Ha, N.-C.; Xu, Y. NADH-dependent butanol dehydrogenase from *Fusobacterium nucleatum*: Purification, crystallization, and X-ray crystallographic analysis. *Biodesign* **2022**, *10*, 29–33. [CrossRef]
43. Otwinowski, Z.; Minor, W. Processing of X-ray diffraction data collected in oscillation mode. *Methods. Enzymol.* **1997**, *276*, 307–326. [PubMed]
44. Wang, Z.; Pan, Q.; Yang, L.; Zhou, H.; Xu, C.; Yu, F.; Wang, Q.; Huang, S.; He, J. Automatic crystal centring procedure at the SSRF macromolecular crystallography beamline. *J. Synchrotron. Radiat.* **2016**, *23 Pt 6*, 1323–1332. [CrossRef] [PubMed]
45. Adams, P.D.; Afonine, P.V.; Bunkoczi, G.; Chen, V.B.; Echols, N.; Headd, J.J.; Hung, L.W.; Jain, S.; Kapral, G.J.; Grosse Kunstleve, R.W.; et al. The Phenix software for automated determination of macromolecular structures. *Methods* **2011**, *55*, 94–106. [CrossRef]
46. Scapin, G. Molecular replacement then and now. *Acta. Crystallogr. D Biol. Crystallogr.* **2013**, *69 Pt 11*, 2266–2275. [CrossRef]
47. Emsley, P.; Lohkamp, B.; Scott, W.G.; Cowtan, K. Features and development of Coot. *Acta. Crystallogr. D Biol. Crystallogr.* **2010**, *66 Pt 4*, 486–501. [CrossRef]
48. Rigsby, R.E.; Parker, A.B. Using the PyMOL application to reinforce visual understanding of protein structure. *Biochem. Mol. Biol. Educ.* **2016**, *44*, 433–437. [CrossRef]
49. Chen, V.B.; Arendall, W.B., 3rd; Headd, J.J.; Keedy, D.A.; Immormino, R.M.; Kapral, G.J.; Murray, L.W.; Richardson, J.S.; Richardson, D.C. MolProbity: All-atom structure validation for macromolecular crystallography. *Acta Crystallogr. D Biol. Crystallogr.* **2010**, *66 Pt 1*, 12–21. [CrossRef]
50. Holm, L.; Rosenstrom, P. Dali server: Conservation mapping in 3D. *Nucleic. Acids. Res.* **2010**, *38*, W545–W549. [CrossRef]
51. Beckmann, L.; Edel, K.H.; Batistic, O.; Kudla, J. A calcium sensor—Protein kinase signaling module diversified in plants and is retained in all lineages of Bikonta species. *Sci. Rep.* **2016**, *6*, 31645. [CrossRef] [PubMed]

International Journal of
Molecular Sciences

MDPI

rticle

Structural Insights and Catalytic Mechanism of 3-Hydroxybutyryl-CoA Dehydrogenase from *Faecalibacterium Prausnitzii* A2-165

aewon Yang [1,†], Hyung Jin Jeon [1,†], Seonha Park [1,‡], Junga Park [1,‡], Seonhye Jang [1], Byeongmin Shin [1], Kyuhyeon Bang [1], Hye-Jin Kim Hawkes [2], Sungha Park [3], Sulhee Kim [1,4] and Kwang Yeon Hwang [1,4,*]

1 Department of Biotechnology, College of Life Sciences and Biotechnology, Korea University, Seoul 02841, Republic of Korea; dkdlemf1475@gmail.com (J.Y.); intiin@hanmail.net (H.J.J.); psh3810@korea.ac.kr (S.P.); jungap@naver.com (J.P.); dbr04198@naver.com (S.J.); kingboom2@naver.com (B.S.); qptmh360@naver.com (K.B.); sulhee@korea.ac.kr (S.K.)

2 Center for Creative Convergence Education, Hanyang University, Seoul 04763, Republic of Korea; hjkhawkes@hanyang.ac.kr

3 Department of Bioengineering, Incheon JEI University, Incheon 21987, Republic of Korea; m00855@jeiu.ac.kr

4 Korea BioDefense Research Institute, Korea University, Seoul 02841, Republic of Korea

* Correspondence: chahong@korea.ac.kr

† These authors are co-first authors.

‡ These authors contributed equally to this work.

Abstract: Atopic dermatitis (AD) is characterized by a T-helper cell type 2 (Th2) inflammatory response leading to skin damage with erythema and edema. Comparative fecal sample analysis has uncovered a strong correlation between AD and *Faecalibacterium prausnitzii* strain A2-165, specifically associated with butyrate production. Therefore, understanding the functional mechanisms of crucial enzymes in the butyrate pathway, such as 3-hydroxybutyryl-CoA dehydrogenase of A2-165 (A2HBD), is imperative. Here, we have successfully elucidated the three-dimensional structure of A2HBD in complex with acetoacetyl-CoA and NAD^+ at a resolution of 2.2Å using the PAL-11C beamline (third generation). Additionally, X-ray data of A2HBD in complex with acetoacetyl-CoA at a resolution of 1.9 Å were collected at PAL-XFEL (fourth generation) utilizing Serial Femtosecond Crystallography (SFX). The monomeric structure of A2HBD consists of two domains, N-terminal and C-terminal, with cofactor binding occurring at the N-terminal domain, while the C-terminal domain facilitates dimerization. Our findings elucidate the binding mode of NAD^+ to A2HBD. Upon acetoacetyl-CoA binding, the crystal structure revealed a significant conformational change in the Clamp-roof domain (root-mean-square deviation of 2.202 Å). Notably, residue R143 plays a critical role in capturing the adenine phosphate ring, underlining its significance in substrate recognition and catalytic activity. The binding mode of acetoacetyl-CoA was also clarified, indicating its lower stability compared to NAD^+. Furthermore, the conformational change of hydrophobic residues near the catalytic cavity upon substrate binding resulted in cavity shrinkage from an open to closed conformation. This study confirms the conformational changes of catalytic triads involved in the catalytic reaction and presents a proposed mechanism for substrate reduction based on structural observations.

Keywords: atopic dermatitis; 3-hydroxybutyryl-CoA dehydrogenase; serial femtosecond crystallography; acetoacetyl-CoA; open–closed conformation

1. Introduction

Atopic dermatitis (AD), also referred to as atopic eczema, is a persistent inflammatory cutaneous disorder typified by pruritic, erythematous, and desquamating skin lesions, and frequently confined to the anatomical regions characterized by flexion [1]. According to the Korean National Health Insurance Service (NHIS), the highest prevalence of AD (26.5%) was observed in children aged 12 to 23 months in 2008, decreasing with age. In addition, the overall prevalence in children under 18 slightly increased from 4.0% in 2003 to 4.5% in 2018, with the prevalence in those aged 6 to 18 rising from 1.9% to 3.1% [2,3]. Clinical manifestations of AD tend to ameliorate with maturation in a subset of affected children. However, approximately half of these children may progress to develop allergic asthma, while approximately two-thirds are at risk of developing allergic rhinitis [3]. In addition to its prevalence in Korea, AD has shown an elevated incidence in other developed nations [4,5]. While environmental factors and genetic predisposition have been considered as primary etiological factors, the rapidity of this phenomenon surpasses any plausible genetic shifts, thereby implicating alterations in the environment associated with Westernized lifestyles as the likely contributing elements [6]. A recent study has shed light on the association between a Westernized diet, characterized by reduced fiber consumption, and the occurrence of distinct microbiome dysbiosis. The intestinal microbiota is extremely plastic in response to diet and environmental factors and, at the same time, governs many aspects of the immune function via short-chain fatty acids (SCFAs) throughout the body [7]. SCFAs are metabolic byproducts originating from the microbiota, comprising a carboxylic acid segment linked to a hydrocarbon tail. SCFAs are mainly produced by microbes in the gut and by skin-commensal bacteria at low concentrations [8]. The significance of SCFAs in AD has been substantiated through research spanning several decades. The gut–skin axis serves as the foundation for systemic allergen sensitization primarily facilitated by the production of SCFAs, notably butyrate. Butyrate enhances skin barrier integrity by modulating the mitochondrial metabolism of epidermal keratinocytes and influencing the synthesis of crucial structural constituents [9,10]. Furthermore, butyrate exerts inhibitory effects on *Staphylococcus aureus*, a microorganism implicated in the upregulation of Th2 (T helper cell 2) cytokines in individuals with AD. This inhibition exacerbates dermal IL-33 expression and skin inflammation via the inhibition of histone deacetylase3 (HDAC3) [11]. Hence, the production of butyrate by the intestinal microbiota exerts a protective influence on AD, contributing to the mitigation of the condition. By the investigation conducted by Han Song [10], a comparative analysis of fecal samples from individuals afflicted with AD and those without the condition revealed a noteworthy reduction in the levels of butyrate, a substance known for its anti-inflammatory properties, in AD patients, and distinctive dysbiosis of microbiota. Pan-genome sequencing of five *Faecalibacterium prausnitzii* strains (A2-165, M21/2, L2-6, S3L/3, and KLE1255) revealed distinctive genetic elements that are responsible for the utilization of significant mucin constituents, such as GalNAc and L-fucose, as well as various nutrients typically released from compromised gut epithelial cells [10]. Notably, the strain L2-6 exhibited a higher count of mapped AD-associated reads compared to non-AD reads, whereas the A2-165 strain displayed the converse pattern [10,12]. Furthermore, the investigation unveiled that dysbiosis within the *F. prausnitzii* species, linked to AD, could potentially lead to the suppression of high butyrate-producing subspecies, notably the A2-165 bacterial type. This was substantiated through a comparative analysis of butyryl CoA: acetate CoA-transferase gene promoter activity among various *F. prausnitzii* strains, suggesting a plausible mechanism resulting in an overall decrease in butyrate production [10,12]. Consequently, comprehending the function of key enzymes such as thiolase, 3-hydroxybutyryl-CoA dehydrogenase, butyryl-CoA dehydrogenase, and butyryl-CoA:acetate CoA transferase in the butyrate pathway is highly significant. Although the structures of these enzymes contributing to the production of various butyrates have been reported [13,14], there have been no comparative studies investigating a difference in production despite the same subspecies. To understand the differences in butyrate production between the L2-6 and A2-165 subspecies, we will aim to clarify the architectures of key enzymes from both subspecies. The A2HBD has elevated expression in *F. prausnitzii* A2-165 and is crucial for the transformation of acetoacetyl-CoA (Figure S1a) to 3-hydroxybutyryl-CoA in the butyrate cycle. We describe the whole structure of A2HBD, including the substrate and co-factor binding modes, as derived from three crystal structures (Figures 1a and 2a,b). Additionally, informed by these structural findings, we propose a putative molecular mechanism for the A2HBD enzyme. In the tertiary-A2HBD structure, the electron density of acetoacetyl-CoA and NAD$^+$ were observed in only one protomer. We propose that this enzyme exhibits cooperativity upon substrate binding, resulting in a conformational change of the cleft. Furthermore, our biochemical findings demonstrated that the binding affinity of acetoacetyl-CoA is less stable than that of NAD$^+$.

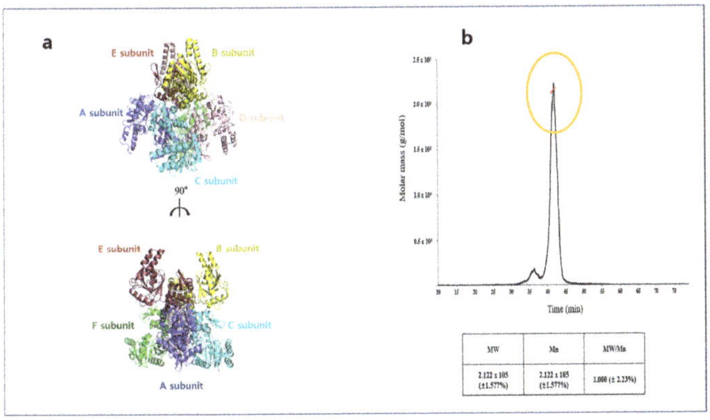

Figure 1. (**a**) Overall structure of apo-A2HBD: Each subunit of hexameric A2HBD is distinguished with different colors (**b**) SEC-MALS of apo-3-Hydroxybutyryl-CoA (A2HBD); sample was measured at 2.018 mg/mL concentration. The protein sample was separated on a size exclusion chromatography (SEC) column. The horizontal line marked as red in yellow circle. represents the measured molar mass. The peak (black) represents the light scattering.

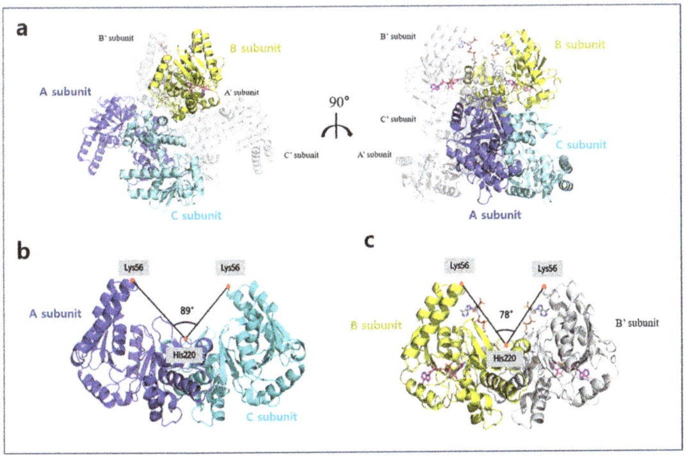

Figure 2. The overall structure of A2HBD in complex with NAD$^+$ and acetoacetyl-CoA. (**a**) The merged hexamer of the A2HBD trimer and the symmetric subunit are shown as a cartoon. The trimeric asymmetric unit of the tertiary complex is shown as a colored cartoon. Subunits A and C are labeled as blue and cyan, respectively. The B subunit of the tertiary complex is shown in yellow. The bound ligand NAD$^+$ and acetoacetyl-CoA are shown in magenta and white, respectively, in a stick model. The symmetric structure of A2HBD is shown in white. (**b**) Comparison of the cleft angle in subunit dimers. Non-bound subunits A and C dimer are shown as cartoons. Subunits A and C are colored blue and cyan, respectively. (**c**) The merged dimer of ligand-bound subunit B and its symmetric unit are shown as cartoons with yellow and white, respectively. The cleft angle is defined by Lys56 at the edge of α2 of the two subunits and His220 at the center of the dimer. The tilted cleft angle is further narrowed by the addition of ligands. Ligands of NAD$^+$ and acetoacetyl-CoA are shown as stick models with magenta and white, respectively.

2. Results

2.1. Overall Structures of Apo-A2HBD Complex with NAD⁺ and Acetoacetyl-CoA

The crystal structure of apo-A2HBD was determined at 2.55 Å resolution. The apo form of A2HBD belonged to $P2_1$ space group with unit cell parameters a = 87.660 Å, b = 126.611 Å, c = 108.071 Å and α = 90 Å, β = 110.208, γ = 90 Å. The apo-A2HBD protein assembles into an asymmetrical hexameric configuration (Figure 1b), wherein all six subunits adopt nearly identical conformations, displaying root-mean-square deviations (RMSD) for the Cα atoms within the range of 0.2 to 0.378 Å (Figure 1a). In the asymmetric unit of crystals, it contains three dimers with three two-fold axes (Figures 1a, 2a and S2c). The crystal structure of the tertiary complex of A2HBD was solved at 2.2 Å resolution (Figure 2a). The tertiary complex belonged to $P3_212$ space group with unit cell parameters a = 90 Å, b = 90 Å, c = 212 Å and α = 90 Å, β = 90, γ = 120 Å. Notably, the tertiary-A2HBD protein manifested as a trimer, wherein only one subunit harbored acetoacetyl-CoA and NAD^+ (Figures S1a,b, 2a and 3a). When the merged hexamer structure (trimer and the crystallographic symmetric unit of the trimer) was superimposed onto the apo-hexamer, the overall structural congruence was strikingly high, with an RMSD of 0.375Å. Furthermore, compared to the dimerization interface observed between the A and C subunits, the B and B′ subunits exhibited a similar dimerization interface (Figure 4). The A2HBD protein exists as a hexamer in solution (Figure 1b). In the tertiary-A2HBD structure, the electron densities of acetoacetyl-CoA and NAD^+ were observed in only one protomer. The exact reason for the complex formation involving only one subunit remains unclear. We propose that, upon substrate binding, the cleft formed between the A, C, and B, B′ subunits contracts by approximately 11 degrees (Figure 2b,c). This hypothesis suggests that this contraction of the dimeric cleft may trigger conformational changes within the hexamer structure, which in turn could affect the solvent accessibility of the other dimer. Such structural alterations might influence the enzyme's substrate preference.

2.2. Monomer of A2HBD

The composition of A2HBD's monomer structure displayed a two-domain topology, featuring distinct N-terminal and C-terminal domains (Figure 3a). The monomer structure consists of eleven α-helices and ten β-strands arranged sequentially, extending from the N-terminus to the C-terminus. Within the N-terminal domain (residues 1–182), a cohesive core structure is formed by the arrangement of seven β-strands, flanked by α-helical regions, among a total of ten β-strands. This core structure shows a typical Rossmann fold pattern, commonly seen in many NAD (P)-dependent dehydrogenases. Similar to a standard Rossmann fold, the first five strands (β1 to β5) in the sheet are arranged side by side, oriented in the same direction. The last two β-strands also exhibit a parallel configuration; however, they extend in the opposite direction compared to the first six strands (Figure 3a). A2HBD also features another unique structural sub-domain known as a helix-turn-helix motif (*Clamp-roof domain*, as herein referenced). This motif (α2, α2- α3, α3) extends from the central Rossmann-fold structure, specifically connecting the β2 and β3 strands. This motif contains several charged and hydrophobic amino acids, including lysine and alanine. Some of these may play a role in interacting with the adenine diphosphate moiety of the acetoacetyl-CoA substrate. Predominantly composed of α-helices, the C-terminal domain (residues 183–288) plays a pivotal role in subunit dimerization (Figure 4). It generates a deep cleft, mainly by α9 and α10 (*Clamp-base domain*), together with the Clamp-roof domain, thereby stabilizing the adenine phosphate moiety of acetoacetyl-CoA (Figures 2b and 3a,b).

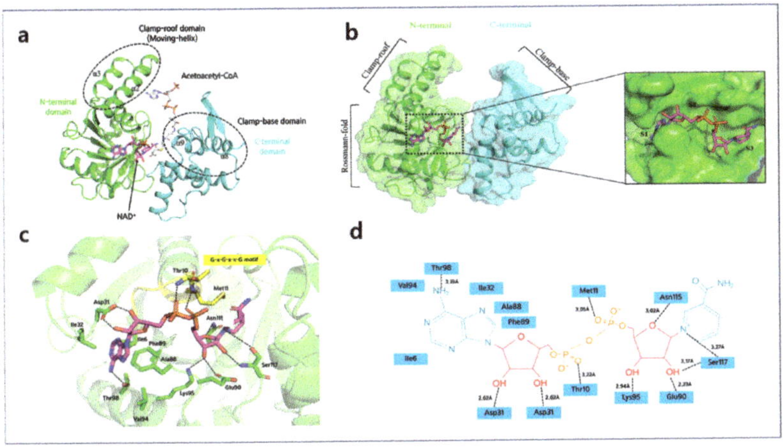

Figure 3. The overall structure of A2HBD in complex with NAD$^+$ and acetoacetyl-CoA. (**a**) Monomer structure of A2HBD complexed with NAD$^+$ and acetoacetyl-CoA. The tertiary monomeric structure is depicted in a cartoon representation, with the N-terminal domain highlighted in green and the C-terminal domain in cyan. The presence of bound NAD$^+$ and acetoacetyl-CoA is illustrated using a stick model, with NAD$^+$ represented in magenta and acetoacetyl-CoA in white. The regions responsible for substrate binding, depicted as the clamp-roof and clamp-base domains, are indicated by a dotted circle in black and are appropriately labeled. (**b**) Crystal structure of A2HBD in complex with NAD$^+$. (left) The "open-book" view of A2HBD monomer is shown in both the cartoon and surface diagram. Three sub-domains and two domains are labeled. The stick diagram of NAD$^+$ is colored magenta. (right) An enlarged view of the cofactor-binding pocket. NAD$^+$ is shown as a stick and A2HBD is shown as a surface. The subsites (S1, S2, S3) of A2HBD are labeled. (**c**) The detailed interaction between NAD$^+$ and A2HBD. The G-x-G-x-x-G motif is marked as yellow. Hydrogen bonding interactions are indicated as black dashed lines. (**d**) Schematic interactions between NAD$^+$ and A2HBD. Hydrogen bonding interactions are indicated as black dashed lines, whereas hydrophobic interactions are not indicated in lines.

Within a single asymmetric unit, two A2HBD protomers—specifically the A and C subunits—engage in mutual contact facilitated by their C-terminal domains, forming a dimeric configuration. The dimerization process is predominantly driven by hydrophobic interactions (Figure 4). The primary contributors to these essential hydrophobic interactions include the $\alpha 9$ and $\alpha 10$ helices, as well as the $\beta 7$ and $\beta 8$ strands. It is worth noting that the hydrophobic residues responsible for mediating this dimerization process remain remarkably conserved among homologous HBD proteins from various species. These residues encompass Val183, Val184, Ile187, Leu188, Ile189, Pro190, Met191, Ile192, Phe197, Ile198, Met200, Val203, Met213, and Leu215. Additionally, the unique antiparallel arrangement of the $\alpha 9$ and $\alpha 10$ helices is further stabilized by the formation of two pairs of salt bridges. Specifically, these salt bridges are formed through interactions between Arg187 and Glu195, as well as Glu202 from one subunit and their counterparts from the other subunit. This Arg/Glu interaction also exhibits high conservation among HBD homologs from diverse species (Figures 4 and S1b).

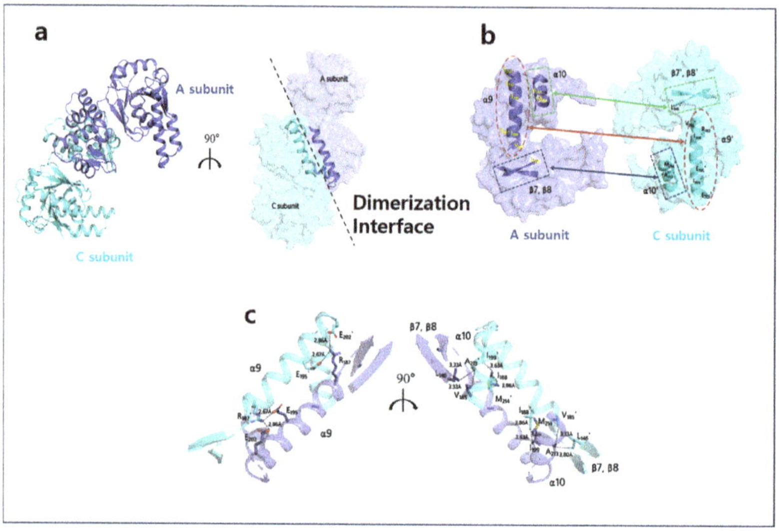

Figure 4. Dimerization interface of A2HBD (**a**) The representative snapshot of the crystal structure of the A2HBD dimer is shown in the cartoon (left) and the surface (right) model. In the cartoon diagram, the subunits A and C are denoted by blue and cyan colors, respectively. They are arranged in a "tail-to-tail" configuration, facilitated by interactions between their C-terminal domains. The surface diagram highlights only the secondary structures relevant to dimerization. (**b**) An "open-book" view of the dimerization interface between subunits A and C. The contact sites of the dimerization interface are indicated with red, green, and blue arrows; and dotted figures. (**c**) Core residues involved in dimerization interface. The combined ribbon and stick model depicts the electrostatic interactions (left) and hydrophobic interactions (right) between the secondary structure elements that contribute to the dimerization interface of the two subunits.

2.3. NAD^+ Binding Site of A2HBD

To ascertain the binding mode of NAD^+, we determined the 2.55Å resolution crystal structure of A2HBD in complex with NAD^+ cofactor (Figure 3a–c). In the structural analysis of the NAD^+ complex derived from crystallography within the $P2_1$ space group, it was observed that all subunits comprising the hexameric assembly were found to bind to NAD^+. By superposing the NAD-bound form of A2HBD with the apo-enzyme, we observed that the monomer structure of the NAD complex is similar to the apo-enzyme structure, displaying a root-mean-square deviation of 0.543Å. Like other NAD-dependent dehydrogenase enzymes, the NAD^+ cofactor binds to the Rossmann fold domain without forming significant interactions with the C-terminal domain (Figure S1a). Observing the binding mode of the NAD cofactor, adenine moiety of NAD cofactor is somewhat exposed to the surface of the enzyme, and the ribose and nicotinamide portion fitted into the deep cleft facing the acetyl group of the substrate. The phosphate moiety of the NAD cofactor resides within the G-x-G-x-x-G motif, composed of Gly7-Ala8-Gly9-Thr10-Met11-Gly12, situated between the initial β strand and the first α a helix within the Rossmann fold domain (Figure S1b). The NAD cofactor is captured by the enzyme's Thr10 and Met11 main-chain nitrogen atoms, establishing hydrogen bonds with the hydroxyl group of the phosphate moiety. Hydrogen bonding interactions involving Asp31, Glu90, Lys95, Asn115, and Ser117, along with hydrophobic interactions with Met11 and Phe138, contribute to the stabilization of the NAD cofactor's two ribose rings and nicotinamide moiety. The hydrophobic pit accommodates the adenine moiety, fostering interactions with hydrophobic residues such as Ile6, Ile32, Ala88, Phe,89, Val94, and Thr98; thus contributing to the stabilization of the adenine phosphate (Figures 3c and S3b).

2.4. Acetoacetyl-CoA Binding Sites of A2HBD

The tertiary complex of HuHAD with NAD^+ and acetoacetyl-CoA demonstrates a notable conformational change upon substrate binding [15]. Specifically, the NAD^+ -binding domain (N-terminal domain) undergoes an inward rotational motion towards the C-terminal domain, facilitating a robust substrate binding interaction [15]. To ascertain if A2HBD exhibits a similar conformational change, we determined the crystal structure of A2HBD in the complex with cofactor and substrate at 2.2 Å (Figures 2a, 3a and 5a). The acetoacetyl-CoA substrate is located between the clamp-roof domain (moving-helix) and clamp-base domain (Figure 5a). Next, the conformational differences of A2HBD with the structures of NAD-complex and tertiary complex were compared. When superposed to the NAD^+ -bound form of the enzyme with the tertiary structure, the overall conformation has been changed, displaying an RMSD of 2.202 Å (Figure 5a,b). Interestingly, overt dissimilarity in structure was conspicuously evident within the N-terminal domain, whereas the C-terminal domain displayed more subtle and discrete conformational alterations. The most notable aspect of this structural change is the movement of the clamp-roof domain, which shifts approximately 6.02 Å closer to the acetoacetyl-CoA compared to its position in the NAD complex, with a rotation of 11 degrees (Figure 5b). This may suggest that when binding of the cofactor, there is no requirement for domain displacement; rather, the transition from an open to a closed conformation plays a pivotal role in facilitating substrate binding (Figure 5c). This hypothesis gains support from the observation that the nicotinamide ring of NAD^+ undergoes an inward movement towards the catalytic cavity during the oxidoreductase reaction with the acetoacetyl moiety of the substrate. The transition from the open to the closed conformation also triggers secondary structural changes, resulting in the formation of the catalytic cavity that contains the catalytic residues. In the open conformation, the cavity's diameter measures approximately 7.5 Å, allowing for ample substrate entry (Figure 5d,e). In contrast, in the closed conformation, this diameter reduces to 5.2 Å, a contraction of 55%. Following substrate binding, the main chain of Asn139 established hydrogen bonds with the substrate, allowing the side chain of Asn139 to align in a manner conducive to Ar-HN interactions (weakly polar interactions between aromatic rings of amino acids and hydrogens of backbone amides) with the aromatic ring of Phe230 (Figure 5d,e). Furthermore, the side chain of Phe230 is closely packed with hydrophobic residues such as Ile231 and Leu227, among others, forming a hydrophobic core near the active site (Figure 5e). In the open conformation, these interactions were disrupted due to alterations in the positioning, geometry, and distances of the residues surrounding Phe230 [16]. It is postulated that the catalytic cavity, shaped by interactions involving Phe230, potentially plays a role in orchestrating the arrangement of catalytic residues, thereby facilitating the occurrence of oxidoreductase activity.

The tertiary structure analysis also elucidated the substrate binding mode of A2HBD. Due to the limited resolution of the electron density map in the adenine moiety of acetoacetyl-CoA obtained from the tertiary structure analysis, a detailed examination of the binding mode between the adenine moiety of the substrate and the enzyme was conducted using fourth-generation PAL-XFEL diffraction data (Figures 6g and S1c). The structures formed almost the same conformation with the RMSD of 0.19Å. Compared to the NAD-binding mode, the acetoacetyl-CoA substrate forms fewer hydrogen bonds with the enzyme (Figure 6e,f), in contrast to the stronger interactions observed with the NAD cofactor, as illustrated in Figure 3d. Therefore, numerous crystallography datasets obtained from the A2HBD, NAD, and acetoacetyl-CoA co-crystals consistently revealed an unclear electron density map for the adenosine phosphate within the substrate (Figure S1c). This suggests a limited level of stabilization for the adenosine ring by the enzyme. This moiety is exposed to the surface, having no binding pocket for the interaction (Figure 5). Instead, Lys54 (NZ) and Lys56 (NZ), located in the clamp-roof domain, form hydrogen bonds with ribose (O48) and O53 of 3′phosphate, respectively. Arg143 (NH1) forms a hydrogen bond with O30 of the substrate within a 3.00 Å distance (Figure 6e,f). Interestingly, in the tertiary structure, rotations of Arg143 toward the adenine ring were observed compared to the NAD complex, allowing the acetoacetyl-CoA to be stabilized through a cation-π interaction.

This stabilization occurs despite the limited involvement of other residues in securing substrate binding (Figure 6e,f). The guanidinium moiety is oriented towards the aromatic ring of the adenine moiety in a stacked (parallel) configuration, with a proximate distance (below 4 Å) between the guanidinium and adenine planes. This arrangement suggests the formation of a cation-π interaction that contributes to the substrate's stabilization [17]. Asn139 mainchain oxygen is positioned to hydrogen bond to the Pantothenic group. The acetoacetyl group is located in hydrogen bonding distance with the sidechains of Ser117 (OB), His136 (NE2), and Asn186 (ND2). These residues interacting with the acetoacetyl moiety are highly conserved (Figure S1b). The hydroxyl group of Ser117 establishes an interaction with the NAD cofactor, thus placing this residue in a bridge with both the substrate and the cofactor. The sidechain of Asn186 located in the C-terminal domain also forms hydrogen bonding with the acetoacetyl moiety of the substrate. Biochemical assays and mutagenesis studies have not yet been undertaken to clarify the primary functions of these residues. Nevertheless, the observations made in this study offer valuable structural insights into the underlying mechanism governing the enzyme's activity.

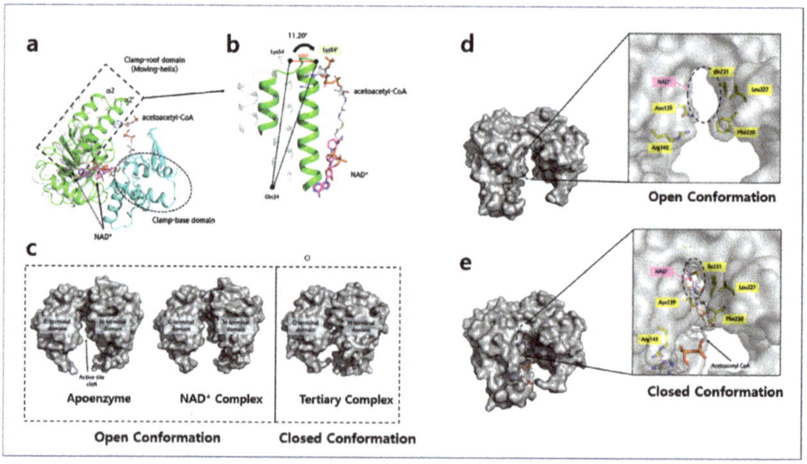

Figure 5. Domain shifting of A2HBD upon acetoacetyl-CoA binding. (**a**) Superimposition of NAD$^+$ complex A2HBD (white) and tertiary complex (green and cyan). Clamp-roof domain movement ($\alpha 2$ to $\alpha 2'$) is described in (**a**) dotted box. NAD$^+$ in cofactor complex, which is transparent, relocated to clear NAD$^+$ (magenta) upon substrate binding. (**b**) The detailed cartoon diagram of clamp-roof domain shifting with (**a**) dihedral angle of 11 degrees upon substrate binding. (**c**) Surface diagram of open-to-closed conformation followed by substrate binding. (**d**,**e**) The open-close conformational change in residues of the catalytic cavity. Amino acid residues involved in forming catalytic cavities are shown as a stick model with a yellow color. "Catalytic cavity" is presented with a dotted circle. Upon substrate binding, the size of the cavity decreased by about 55%. The cation-π interaction (Asn139-Phe230) is shown as dotted red lines.

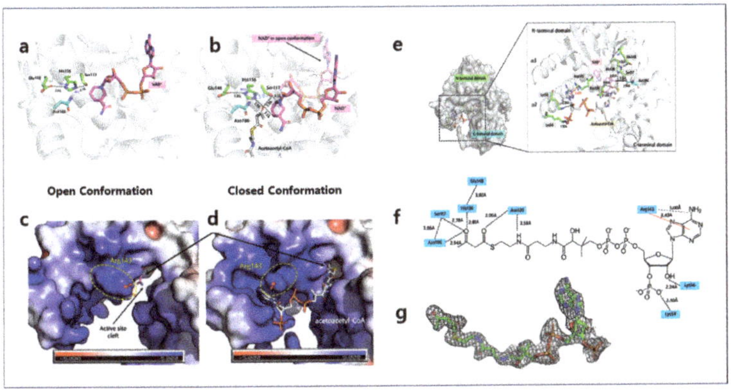

Figure 6. (**a**,**b**) The open-closed conformational change of interactions within catalytic residues in A2HBD upon substrate binding. In both panels, A2HBD structures are shown as gray-colored cartoons. The catalytic residues (Ser117, His136, Glu148) and NAD^+ are shown as stick diagrams with green and magenta colors, respectively. (**a**) The open conformation occurs when only the NAD^+ cofactor is bound. The hydrogen bonding interactions between catalytic residues are presented as dotted lines. Three catalytic residues are bridged with hydrogen bonding interactions. (**b**) Upon substrate binding, the ligands and the catalytic residues adopt closed conformation. The hydrogen bonding interactions are marked as dotted lines. The distance between the NE2 of His136 and the side chain oxygen of Ser117 residue is presented with a pale-blue-colored dotted line labeled as 4.9Å. The distance became more distant from 2.7Å to 4.9Å in the transition to closed conformation, indicating breakage of hydrogen bonding. (**c**,**d**) Surface electrostatic presentation of the substrate binding site; A2HBD structures without (**c**) and with (**d**) substrate binding are presented as surface electrostatic presentations with the same orientation. The arginine 143 residue is marked as a yellow dotted circle and labeled. The black arrow indicates substrate binding pocket. The orientation of the red-colored arrow indicates the downside of the guanidinium side chain of arginine. Following substrate binding, the guanidinium side chain rotates approximately 90 degrees to align parallel to the adenine group of acetoacetyl-CoA to stabilize the substrate binding. (**e**–**g**) Tertiary crystal structure of A2HBD in complex with NAD^+ and acetoacetyl-CoA. (**e**) Overall tertiary complex of A2HBD. A2HBD is shown as a gray surface model. The acetoacetyl-CoA and substrate-interacting residues are shown as stick diagrams. (**f**) The acetoacetyl-CoA is presented as white and N-terminal residues and C-terminal residues interacting with the substrate. Hydrogen bonding interactions are presented as dotted lines and cation-π interaction is represented as red-colored dotted lines. (**g**) The 2fofc electron density map (1.0 σ) of acetoacetyl-CoA by SFX-data.

3. Discussion

The structural analysis of the NAD^+ and tertiary complex structures reveals that the catalytic residues, specifically Ser117, His136, and Glu148, are positioned near the active site. This domain encompasses both the acetyl group of the substrate and the amide group of the NAD^+ nicotinamide, suggesting their essential roles in the catalytic process (Figures S3a and 7a). These residues are part of the conserved sequence, HXFX-PXXXMXLXE, which shows strong conservation across the HBD families, as noted in previous studies [18–20]. The molecular mechanism underlying HBD catalysis involving this conserved catalytic domain has been thoroughly investigated in prior research [17–20]. The validity of these mechanisms has been substantiated through the analysis of the crystal structure of the A2HBD tertiary complex. The structures revealed that Ser117, His136, and Glu148 constitute the catalytic triad, engaged in hydrogen bonding interactions within the NAD^+ complex (Figure S3a). When comparing the catalytic triad's conformational changes from the open to closed conformations (Figures 5c and 7a), it becomes evident that in the NAD^+ complex, characterized by an open conformation, the catalytic triad resides within a hydrogen bonding distance from each other, thereby contributing to the stabilization of these critical residues (Figure S3a). In the crystallization solution, assessed at a pH range of 5.3 to 5.6, it was observed that the imidazole ring of His136 exists in a protonated state, while the side chain of Glu148 remains negatively charged. This finding indicates that the protonated state of His136 is stabilized by the negative charge present on Glu148, which is referred to as the His/Glu pair, a phenomenon also supported by observations in the case of human 3-hydroxyacyl-CoA dehydrogenase [21]. After substrate binding, NAD^+ undergoes an inward movement, positioning the nicotinamide moiety within the active site. Simultaneously, the acetoacetyl group of the substrate is situated adjacent to the nicotinamide moiety within the active site. These specific placements result in Ser117 being positioned beyond the hydrogen bonding distance with His136, measured at approximately 4.9 Å.

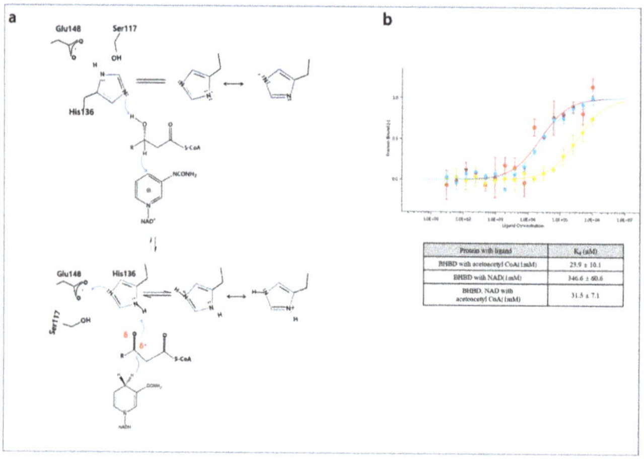

Figure 7. (**a**) Structure-based catalytic mechanism of the acetoacetyl-CoA-catalyzed reaction. Schematic diagram of proposed catalytic mechanism in A2HBD active site. Hydride transfer and nucleophilic substitution are presented as blue-colored arrows. Polarized carbonyl carbon and oxygen are indicated as red-colored δ^+ and δ^-', respectively. (**b**) Binding affinity analysis of NAD^+, acetoacetyl-CoA, and NAD^+/acetoacetyl-CoA by MST experiments. The K_d values were showed BHBD with acetoacetyl-CoA in red color; BHBD in NAD^+, and acetoacetyl-CoA in cyan color; BHBD with NAD^+ in yellow color, respectively. Each K_d value is shown in the table.

However, the His136 and Glu148 maintain hydrogen bonding interactions, with His136 notably forming a bond with the O3 atom of the substrate at a distance of approximately 2.9 Å (Figures 6b and 7a). The disruption of hydrogen bonding between Ser117 and His136 in a substrate-dependent manner suggests a potential role for His136 as a general acid/base catalyst and sheds light on the mechanism of nucleophilic reactivity within the enzyme (Figure S3a). Under the modulation of the catalytic triad through the open-closed conformational transition, we could confirm and propose a conserved mechanism for acetoacetyl-CoA-catalyzed reactions in A2HBD, as shown in Figure 7a. Upon substrate binding, acetoacetyl-CoA is stabilized in the closed conformation through interactions with the conserved active site residues (Figures 6e and S1b). The positive charge carried by the protonated imidazole ring of His136 plays a crucial role in stabilizing the partial negative charge on the C3 oxygen (O3) within the transition state of the substrate. Additionally, it aids in polarizing the carbonyl moiety of the substrate. Following this polarization, the carbonyl group develops an electron-deficient carbon, facilitating the transfer of a hydride ion from NADH to the C3 carbon. Subsequently, His136 functions as a proton donor, leading to the reduction of acetoacetyl-CoA to 3-hydroxybutyryl-CoA. This deprotonation event disrupts the hydrogen bonding interaction with Ser117 (Figures S3a and 7a). The binding affinity of acetoacetyl-CoA to A2HBD is 23.9 μM while that of the NAD^+ is 346.6 μM (Figure 7b). In the case of NAD^+, its binding affinity for A2HBD (31.5 μM) is stronger when the substrate (acetoacetyl-CoA) is present, compared to when it is alone (Figure 7b). We may suggest that this is related to the formation of the tertiary complex in only one monomer within the trimer, which may help explain the cooperative nature of this enzyme (Figure 2a). While the structural data from A2HBD supports the proposed catalytic mechanism, further confirmation of the roles played by the catalytic residues necessitates site-directed mutagenesis and enzymatic assays in future investigations.

4. Materials and Methods

4.1. Cloning and Protein Expression

The A2-165 BHBD coding gene (Met1-Leu290) with a calculated molecular weight of 31.1 kDa, underwent enzymatic amplification via polymerase chain reaction (PCR) utilizing chromosomal DNA isolated from *F. praunsnitzii* as the template. Subsequently, the resulting PCR product of the A2-165 strain (DSM 17677) was integrated into a $pET21a^+$ vector (Novagen, Sigma-Aldrich, Burlington, MA, USA) containing a 6x histidine tag at the C-terminus. Amplification of gene fragments was achieved employing primers 5′-**ctagctagc**atgaagatcggtgttatc-3′ (A2HBD_forward) and 5′-**ccgctcgag**ctggtcaacaggggtc-3′ (A2HBD_reverse), resulting in the amplification of the full-length 867bp A2-165 BHBD gene. Notably, the bold sequence section in the primer design corresponds to the incorporation of *NheI* and *XhoI* restriction enzyme recognition sites at the 5′-end of the forward and reverse primers, respectively. The constructed expression vector was then introduced into *Escherichia coli* BL21 (DE3) for protein expression. The transformed cells were cultured in 4L of LB medium containing ampicillin (50 mg/mL) at 37 °C, until reaching an absorbance 0.6~0.8 at OD_{600}. Protein production was induced by 1.0mM Isopropyl β-D-1-thiogalactopyranoside (IPTG), and then the cells were incubated for 18 h at 18 °C. The grown cells were harvested by centrifugation at 1600× *g* at 277.15 K for 1 h. The cell pellets were resuspended in lysis buffer (20 mM Tris-HCl at pH 8.0 and 150 mM NaCl) and then disrupted by ultrasonication.

4.2. Protein Production and Crystallization

The cell debris was removed by centrifugation at 13,000 rpm for 60 min. The supernatant was filtered through a 0.45 μm polyvinylidene difluoride (PVDF) membrane filter (Merck Millipore, Seoul, Republic of Korea) and was bound to a 5 mL His Trap HP column (GE Healthcare, Seoul, Republic of Korea). After washing with His binding buffer (20 mM Tris-HCl at pH 8.0 and 150 mM NaCl), the bound protein was then eluted with His binding buffer containing 500 mM imidazole in a linear gradient. To eliminate trace contaminants and facilitate crystallization, the protein sample was subjected to size-exclusion chromatography using a HiLoad 16/600 Superdex 75 prep-grade column (GE Healthcare). The column was pre-equilibrated with a final buffer solution consisting of 20 mM Tris-HCl at pH 8.0 and 150 mM NaCl. Following chromatography, the purified protein, which exhibited an approximate purity of 95%, as determined by SDS-PAGE analysis, was concentrated to a final concentration of 53.3 mg/mL. This concentration step was achieved using an Amicon filter (Merck, Seoul, Republic of Korea) with the final buffer as the processing medium.

The initial screening for the crystallization was performed by the sitting-drop vapor diffusion method using the commercial crystallization screening kit (Hampton Research) in MRC 2-well Crystallization Plates (Hampton Research, Seoul, Republic of Korea) at 293 K. MCSG crystal screening solution (Anatrace, Maumee, OH, USA) was used as a screening solution. A volume of 60 μL of the screening solution was dispensed into each reservoir well, and the screening process was conducted by combining 0.5 μL of the screening solution with an equivalent amount of protein sample. After a 14-day incubation period at a temperature of 293 K, we successfully generated crystals suitable for diffraction experiments. These crystals were formed through co-crystallization of the substrate, cofactor, and tertiary complex using the same method and screening conditions. Notably, the crystallization process was supplemented with 5 mM each of NAD^+ and acetoacetyl-CoA. The crystal solution conditions were further optimized in a hanging-drop vapor diffusion method adjusting the pH, salt, and precipitant concentration. 1 µL of 20 mg/mL protein solution was mixed with precipitation solution at a ratio of 1:1 in 24 well VDX plates (Hampton Research) at 293 K. The method employed for creating the complex crystal was identical, except for the omission of 5 mM each of NAD^+ and acetoacetyl-CoA. After going through several steps for optimization, the best quality of crystals and the maximum size of macro-crystals appeared after 14 days in a solution condition of 0.23 M Ammonium citrate dibasic (pH 5.0) and 1 5% PEG3350 for the apo- and tertiary complex. For NAD^+ complex, the solution condition of 0.25 M Lithium Sulfate, 0.1 M Tris-HCl, and 1.28 M Ammonium Sulfate produced the best-quality crystals [22].

In pursuit of generating a substantial quantity of micro-crystals for serial femtosecond crystallography (SFX) experiments, which necessitate the acquisition of diffraction data from a multitude of micro-crystals possessing random orientations [23], a crystallization process was initiated via batch-crystallization [24]. Before crystallization, 20 mg/mL of A2HBD protein solution was incubated with 5 mM acetoacetyl-CoA and 5mM NAD^+ cofactor for 30 min. In the initial batch experiment, we amalgamated 10 µL of protein solution with a concentration of 20 mg/mL and a precipitant solution comprising 0.21 M to 0.28 M Ammonium citrate dibasic (pH 5.0) along with 15% PEG 3350, utilizing ratios of 1:1, 1:2, and 1:3, respectively. The resulting mixture was subjected to gentle mixing within Eppendorf tubes (EP tubes) using a pipette, followed by incubation in sealed EP tubes at 293 K to prevent evaporation. Notably, during this step, meticulous care was taken to ensure a slow addition of the precipitant solution to the protein, with a duration of 2–5 s, to avert the formation of localized high concentrations of precipitants [24]. Following the 5~7 days incubation period, crystals with dimensions less than 0.2 mm were successfully obtained in the mixture solution of a ratio of 1:3.

4.3. Data Collection for Apo, NAD^+ Complex and Tertiary Comple

Before data collection at 100 K, the crystals were transferred to cryoprotectant solution containing 0.23 M ammonium citrate dibasic (pH 5.0), 15% PEG3350 and 25% glycerol. The crystals were flash-frozen in liquid nitrogen and then processed into a cryoprotectant solution containing 0.23 M ammonium citrate dibasic (pH 5.0), along with a mixture of 15% PEG3350 and 25% glycerol. For NAD^+ co-incubated crystals, the cryoprotectant solution included 0.25 M lithium sulfate, 0.1 M Tris/HCl, 1.28 M ammonium sulfate and 25% 2-Methyl-2,4-pentanediol [22]. Additionally, 20 mM of cofactor and substrate-cofactor were introduced into the mixture. Following this, the crystals were promptly flash-frozen by immersion in liquid nitrogen. The X-ray diffraction data were collected at Pohang Accelerator Laboratory (PAL) beamline 11C. The diffraction frames were obtained on the PILATUS3 X 6 M detector using 1° rotation and 1s exposure time for 360°. The data set was indexed, integrated, and scaled with the HKL2000 suite [25].

The X-ray diffraction data of the acetoacetyl-CoA complex were collected at the Pohang Accelerator Laboratory X-ray Free Electron Laser (PAL-XFEL) facility. The fixed-sample scanning method was utilized to precisely position and transport the crystal sample to the designated X-ray interaction point [26]. For the sample delivery system, a Nylon-Mesh sample holder was employed. This holder, composed of Nylon-Mesh containing crystals enveloped by polyimide films, offered the advantage of minimal background scattering when subjected to X-rays. This characteristic allowed us to circumvent issues related to dehydration and enabled the acquisition of highly precise diffraction data for subsequent analysis [27]. Prior to mounting, the crystal solutions were pipetted to fragment the crystals into sizes below 200 µm, ensuring they could be effectively positioned on the pores of the Nylon-Mesh sample holder. The sample was then mounted inside the FT-SFX chamber equipped with helium purging environments. The diffraction images were captured using a Rayonix NX225-HS detector. Using a Cheetah software package (version 1.12), the data were indexed, integrated, and scaled [28].

4.4. Structure Determination

In the process of deriving a model structure for the protein crystals, the Python-based Hierarchical Environment for Integrated Xtallography (Phenix) software program (version 1.20-4459) was employed [29,30]. This involved inputting a protein sequence, a data file, and a reference model. Initially, we solved the structure through molecular replacement (MR) using the 3-hydroxybutyryl-CoA dehydrogenase from *Clostridium butylicum* (PDB ID: 4KUE) as the reference model, utilizing the Phaser module within PHENIX [29]. Subsequently, 90% of the residues were automatically built using the Autobuild module in PHENIX [30]. Further refinement and manual model building were conducted with the assistance of the Crystallographic Object-Oriented Toolkit (Coot 0.9.8.95) [31]. The model structure was fitted with an electron-density map using Coot 0.9.8.95, and final refinement was executed using the Refine module within PHENIX. Further model building and refinement were performed using REFMACS in CCP4 (version 0.9.0.003) and PHENIX (version 1.21). The final model for apo-A2HBD had R values of R_{factor} = 21.5%, R_{free} = 25.7% at 2.75 Å resolution; NAD^+/A2HBD had R values of R_{factor} = 20.6%, R_{free} = 26.6% at 2.55 Å resolution; acetoacetyl-CoA/A2HBD (S117A) had R values of R_{factor} = 18.0%, R_{free} = 22.2% at 1.90 Å resolution; NAD^+/acetoacetyl-CoA/A2HBD (S117A) had R values of R_{factor} = 19.9%, R_{free} = 24.4% at 2.2 Å resolution. Finally, the structural figures were generated using PyMOL 3.0 (Educational version) [32]. The statistics for the detailed data collection and structure refinement are provided in Table 1.

Table 1. Data collection and refinement statistics for apo- and complex structures of 3-Hydroxybutyryl-CoA dehydrogenase.

	Apo	Native with NAD	S117A with AACoA	S117A with AACoA&NAD
Data collection				
X-ray source	PAL11C	PAL11C	SFX	PAL11C
Wavelength (Å)	1.000	1.000	1.000	1.000
Space group	$P2_1$	$P2_1$	$P3_212$	$P3_212$
a, b, c (Å)	87.66, 126.61, 108.09	90.82, 80.48, 128.35	91.01, 91.01, 212.7	90.56, 90.56,212.50
α, β, γ (°)	90.0, 110.2, 90.0	90.0, 102.2, 90.0	90, 90, 120	90, 90, 120
Resolution (Å)	48.04–2.69 (2.72–2.69)	46.64–2.55 (2.58–2.55)	45.51–1.90 (1.97–1.90)	45.28–2.20 (2.23–2.2)
Unique reflections	60,712 (1807)	54,530 (1522)	190,068 (7874)	50,509 (1392)
Completeness	96.7 (78.3)	91.7 (77.2)	99.99 (100.0)	99.0 (83.6)
Redundancy	6.8 (5.5)	4.1 (2.3)	3794 (189.7)	10.2 (3.8)
I/σ	10	23.1	26.1	15.6
Rmerge (%)	0.22 (0.75)	0.08 (0.25)	0.09 (0.26)	0.19 (0.70)
Refinement statistics				
Resolution (A)	48.04–2.69	46.64–2.55	45.51–1.9	45.28–2.20
Reflections used in refinement	60,712	54,530	190,068	50,509
Rwork/Rfree	0.2458/0.2943	0.2058/0.2661	0.1795/0.2224	0.1988/0.2442
R.M.S. deviations				
Bond lengths (A)	0.011	0.017	0.010	0.022
Bond angles (°)	1.18	1.24	1.26	1.4
Ramachandran plot (%)				
Favored	92.28	93.94	96.26	99.06
Allowed	6.16	5.65	3.51	0.94
No. atoms				
Protein	12,930	12,939	6423	6423
Ligands	0	264	98	98
Solvent	0	79	250	250
B-factors	52	62.3	26.2	37.4

Statistics for the highest resolution shell are shown in parentheses.

4.5. Multi-Angle Light Scattering (MALS) Analysis

Multi-angle light scattering (MALS) was conducted to determine the absolute molecular weight of the A2HBD. The AKTA chromatography system was integrated with a MALS detector TREOS II MALS (WYATT). The purified protein concentration was diluted in the final buffer (20 mM Tris-HCl at pH 8.0 and 150 mM NaCl) to 2.018 mg/mL and analyzed using SEC-MALS. Protein solution was loaded on a Superdex 200 Increase analytical column (GE Healthcare), which is pre-equilibrated in a final buffer. The buffer's flow rate was consistently upheld at 0.3 mL/min. Before experimenting, the bovine serum albumin (BSA) was quantified as a baseline reference measurement. ASTRA software (https://wyatt.com/products/software/astra.html, accessed on 30 September 2024, Wyatt Technology, Goleta, CA, USA), provided by MALS manufacturer, was utilized to process and analyze the data. The elution profile of A2HBD exhibited a predominant peak, which accounted for 92% of the scattered mass and was estimated to be approximately 212.2 kDa with a precision of $\pm$1.577%. Considering its monomer molecular weight of 32.4 kDa, this mass value provides the estimate for a hexameric state when A2HBD is present in the solution (Figure 1b).

4.6. MicroScale Thermophoresis (MST) Measurement

MST measurement was conducted to identify molecular interaction between A2HBD and substrate or cofactor. A2BHD protein was labeled fluorescently using a His-tag Labeling Kit (The Monolith His-Tag Labeling Kit RED-tris-NTA 2nd generation: Nanotemper technologies, South San Francisco, CA, USA) at 50 nM concentration. We started at 0.55 mM, and 0.2 mM, respectively. The proteins were serially diluted to 16 concentrations. 1:1 protein–protein solutions were incubated for 30 min. The measurements were conducted under conditions where the LED power and MST power were both at 40%. The affinity constant (Kd) was determined by curve fitting using MO Affinity Analysis software v2.1.3 (Nanotemper, South San Francisco, CA, USA). All measurements were conducted in triplicate (Figure 7b).

5. Conclusions

Some scientists have been studying the link between butyrate metabolism and atopic dermatitis (AD) [10,12]. To understand the butyrate metabolism pathway, we have determined the four crystal structures of 3-hydroxybutyryl-CoA dehydrogenase from *Faecalibacterium prausnitzii* strain A2-165 (A2HBD) in the presence of acetoacetyl-CoA and NAD^+. This was achieved using the PAL-11C beamline (a third-generation facility) and PAL-XFEL (a fourth-generation facility) with the technique of serial femtosecond crystallography (SFX). In the tertiary-A2HBD structure, the electron density of acetoacetyl-CoA and NAD^+ were observed in only one protomer. We propose that this enzyme exhibits cooperativity upon substrate binding, resulting in a conformational change of the cleft. Furthermore, our biochemical findings demonstrated that the binding affinity of acetoacetyl-CoA is less stable than that of NAD^+. The Arg143 has a critical role in identifying substrates and facilitating catalytic activity. The binding mechanism of acetoacetyl-CoA was studied further, revealing that it is less stable than NAD^+. This work validates the conformational changes of catalytic triads involved in the catalytic reaction. Furthermore, it presents a theorized mechanism for substrate reduction based on structural discoveries. Our biochemical findings demonstrated that the binding affinity of acetoacetyl-CoA is less stable than that of NAD^+.

Supplementary Materials: The following supporting information can be downloaded at: https://www.mdpi.com/article/10.3390/ijms251910711/s1.

Author Contributions: Conceptualization, K.Y.H.; methodology, J.Y., H.J.J., J.P., B.S., K.B., S.K., S.P. (Sungha Park) and K.Y.H.; software, J.Y.; S.J. and S.K.; validation, J.Y., S.P. (Seonha Park) and S.K.; formal analysis, S.K., S.P. (Seonha Park) and J.Y.; data curation, J.Y.; H.J.J., S.K., S.P. (Sungha Park) and S.P. (Seonha Park); writing—original draft preparation, K.Y.H., S.P. (Seonha Park), H.-J.K.H. and K.Y.H.; writing—review and editing, H.J.J., H.-J.K.H., K.Y.H., S.P. (Seonha Park) and K.Y.H.;

visualization, S.K., H.J.J., S.P. (Seonha Park) and K.Y.H.; supervision, K.Y.H.; project administration, K.Y.H.; funding acquisition, S.K. and K.Y.H. All authors have read and agreed to the published version of the manuscript.

Funding: This work was supported by the National Research Foundation of Korea (NRF) grants funded by the Korean government (MSIT) (NRF-2020R1A2C2005670 to K.Y.H. NRF-2022R1I1A1A01068223 to S.K.). S.K. and K.Y.H. are supported by Korea University Grants.

Institutional Review Board Statement: Not applicable.

Informed Consent Statement: Not applicable.

Data Availability Statement: The accession numbers for the structures of *Apo-A2HBD, A2HBD complex with NAD +, A2HBD (S117A) complex with acetoacetyl-CoA*, and *A2HBD (S117A) complex with NAD +/acetoacetyl-CoA* as described in this paper, are PDB9JIO, 9JHE, 9JHY, and 9JHZ, respectively. All other data are included in this paper.

Acknowledgments: We thank the research staff at Korea Basic Science Institute and K.H. Nam for XFEL processing, PAL11C, and PAL-XFEL (S.J. Lee & J. Park) of Pohang Light Source in Republic of Korea for the use of their excellent facilities and assistance. Diffraction data were collected at BL44XU of the Spring-8 in Japan facility under Proposals 2023A6675 and 2024A6975. K.B. and B.S. are supported by NRF-RS-2022-00143178.

Conflicts of Interest: The authors declare no conflicts of interest.

References

1. Berke, R.; Singh, A.; Guralnick, M. Atopic dermatitis: An overview. *Am. Fam. Physician* **2012**, *86*, 35–42. [PubMed]
2. Yu, J.-S.; Lee, C.-J.; Lee, H.-S.; Kim, J.; Han, Y.; Ahn, K.; Lee, S.-I. Prevalence of atopic dermatitis in Korea: Analysis by using national statistics. *J. Korean Med. Sci.* **2012**, *27*, 681. [CrossRef]
3. Ricci, G.; Patrizi, A.; Baldi, E.; Menna, G.; Tabanelli, M.; Masi, M. Long-term follow-up of atopic dermatitis: Retrospective analysis of related risk factors and association with concomitant allergic diseases. *J. Am. Acad. Dermatol.* **2006**, *55*, 765–771. [CrossRef] [PubMed]
4. Beasley, R.; The International Study of Asthma and Allergies in Childhood Steering Committee. Worldwide variation in prevalence of symptoms of asthma allergic rhinoconjunctivitis, and atopic eczema: ISAAC. *Lancet* **1998**, *351*, 1225–1232. [CrossRef]
5. Lee, J.Y.; Kim, J.; Ahn, K. Time trends in the prevalence of atopic dermatitis in Korean children according to age. *Allergy Asthma Immunol. Res.* **2022**, *14*, 123. [CrossRef] [PubMed]
6. Ehlers, S.; Kaufmann, S.H. Infection, inflammation, and chronic diseases: Consequences of a modern lifestyle. *Trends Immunol.* **2010**, *31*, 184–190. [CrossRef]
7. Lee, Y.K.; Mazmanian, S.K. Has the microbiota played a critical role in the evolution of the adaptive immune system? *Science* **2010**, *330*, 1768–1773. [CrossRef]
8. Furusawa, Y.; Obata, Y.; Fukuda, S.; Endo, T.A.; Nakato, G.; Takahashi, D.; Nakanishi, Y.; Uetake, C.; Kato, K.; Kato, T. Commensal microbe-derived butyrate induces the differentiation of colonic regulatory T cells. *Nature* **2013**, *504*, 446–450. [CrossRef]
9. Trompette, A.; Pernot, J.; Perdijk, O.; Alqahtani, R.A.A.; Santo Domingo, J.; Camacho-Muñoz, D.; Wong, N.C.; Kendall, A.C.; Wiederkehr, A.; Nicod, L.P. Gut-derived short-chain fatty acids modulate skin barrier integrity by promoting keratinocyte metabolism and differentiation. *Mucosal Immunol.* **2022**, *15*, 908–926. [CrossRef]
10. Song, H.; Yoo, Y.; Hwang, J.; Na, Y.-C.; Kim, H.S. *Faecalibacterium prausnitzii* subspecies–level dysbiosis in the human gut microbiome underlying atopic dermatitis. *J. Allergy Clin. Immunol.* **2016**, *137*, 852–860. [CrossRef]
11. Luo, C.-H.; Lai, A.C.-Y.; Chang, Y.-J. Butyrate inhibits Staphylococcus aureus-aggravated dermal IL-33 expression and skin inflammation through histone deacetylase inhibition. *Front. Immunol.* **2023**, *14*, 1114699. [CrossRef] [PubMed]
12. Kiefer, D.; Ali-Akbarian, L. A brief evidence-based review of two gastrointestinal illnesses: Irritable bowel and leaky gut syndromes. *Integr. Med.* **2004**, *3*, 1–11.
13. Takenoya, M.; Taguchi, S.; Yajima, S. Crystal structure and kinetic analyses of a hexameric form of (S)-3-hydroxybutyryl-CoA dehydrogenase from Clostridium acetobutylicum. *Acta Crystallogr. Sect. F Struct. Biol. Commun.* **2018**, *74*, 733–740. [CrossRef] [PubMed]
14. Kim, E.-J.; Kim, J.; Ahn, J.-W.; Kim, Y.-J.; Chang, J.H.; Kim, K.-J. Crystal structure of (S)-3-hydroxybutyryl-CoA dehydrogenase from Clostridium butyricum and its mutations that enhance reaction kinetics. *J. Microbiol. Biotechnol.* **2014**, *24*, 1636–1643. [CrossRef] [PubMed]
15. Barycki, J.J.; O'Brien, L.K.; Bratt, J.M.; Zhang, R.; Sanishvili, R.; Strauss, A.W.; Banaszak, L.J. Biochemical characterization and crystal structure determination of human heart short chain L-3-hydroxyacyl-CoA dehydrogenase provide insights into catalytic mechanism. *Biochemistry* **1999**, *38*, 5786–5798. [CrossRef] [PubMed]
16. Tóth, G.; Watts, C.R.; Murphy, R.F.; Lovas, S. Significance of aromatic-backbone amide interactions in protein structure. *Proteins Struct. Funct. Bioinform.* **2001**, *43*, 373–381. [CrossRef]

17. Infield, D.T.; Rasouli, A.; Galles, G.D.; Chipot, C.; Tajkhorshid, E.; Ahern, C.A. Cation-π interactions and their functional roles in membrane proteins. *J. Mol. Biol.* **2021**, *433*, 167035. [CrossRef]
18. Yang, S.-Y.; He, X.-Y. Molecular mechanisms of fatty acid β-oxidation enzyme catalysis. In *Current Views of Fatty Acid Oxidation and Ketogenesis: From Organelles to Point Mutations*; Quant, P.A., Eaton, S., Eds.; Springer: Boston, MA, USA, 1999; pp. 133–143.
19. He, X.-Y.; Zhang, G.; Blecha, F.; Yang, S.-Y. Identity of heart and liver L-3-hydroxyacyl coenzyme A dehydrogenase. *Biochim. Biophys. Acta (BBA)-Mol. Cell Biol. Lipids* **1999**, *1437*, 119–123. [CrossRef]
20. Barycki, J.J.; O'BRIEN, L.K.; Birktoft, J.J.; Strauss, A.W.; Banaszak, L.J. Pig heart short chain L-3-hydroxyacyl-CoA dehydrogenase revisited: Sequence analysis and crystal structure determination. *Protein Sci.* **1999**, *8*, 2010–2018. [CrossRef]
21. Barycki, J.J.; O'Brien, L.K.; Strauss, A.W.; Banaszak, L.J. Glutamate 170 of human l-3-hydroxyacyl-CoA dehydrogenase is required for proper orientation of the catalytic histidine and structural integrity of the enzyme. *J. Biol. Chem.* **2001**, *276*, 36718–36726. [CrossRef]
22. Park, J. Structural Analysis of 3-Hydroxybutyryl-CoA Dehyrogenase from *Faecalibacterium prausnitzii* A2-165. Master's Thesis, Korea University, Seoul, Republic of Korea, 2021.
23. Boutet, S.; Lomb, L.; Williams, G.J.; Barends, T.R.; Aquila, A.; Doak, R.B.; Weierstall, U.; DePonte, D.P.; Steinbrener, J.; Shoeman, R.L. High-resolution protein structure determination by serial femtosecond crystallography. *Science* **2012**, *337*, 362–364. [CrossRef] [PubMed]
24. Rayment, I. Small-scale batch crystallization of proteins revisited: An underutilized way to grow large protein crystals. *Structure* **2002**, *10*, 147–151. [CrossRef] [PubMed]
25. ZaM, W.O. Processing of X-ray diffraction data collected in oscillation mode. *Methods Enzymol.* **1997**, *276 Pt A*, 20.
26. Martiel, I.; Müller-Werkmeister, H.M.; Cohen, A.E. Strategies for sample delivery for femtosecond crystallography. *Biol. Crystallogr.* **2019**, *75*, 160–177. [CrossRef] [PubMed]
27. Park, J.; Nam, K.H. Sample delivery systems for serial femtosecond crystallography at the PAL-XFEL. *Photonics* **2023**, *10*, 557. [CrossRef]
28. Barty, A.; Kirian, R.A.; Maia, F.R.; Hantke, M.; Yoon, C.H.; White, T.A.; Chapman, H. Cheetah: Software for high-throughput reduction and analysis of serial femtosecond X-ray diffraction data. *J. Appl. Crystallogr.* **2014**, *47*, 1118–1131. [CrossRef]
29. McCoy, A.J.; Grosse-Kunstleve, R.W.; Adams, P.D.; Winn, M.D.; Storoni, L.C.; Read, R.J. Phaser crystallographic software. *J. Appl. Crystallogr.* **2007**, *40*, 658–674. [CrossRef]
30. Afonine, P.V.; Grosse-Kunstleve, R.W.; Echols, N.; Headd, J.J.; Moriarty, N.W.; Mustyakimov, M.; Terwilliger, T.C.; Urzhumtsev, A.; Zwart, P.H.; Adams, P.D. Towards automated crystallographic structure refinement with phenix. refine. *Acta Crystallogr. Sect. D Biol. Crystallogr.* **2012**, *68*, 352–367. [CrossRef]
31. Emsley, P.; Cowtan, K. Coot: Model-building tools for molecular graphics. *Acta Crystallogr. Sect. D Biol. Crystallogr.* **2004**, *60*, 2126–2132. [CrossRef]
32. DeLano, W.L.; Lam, J.W. PyMOL: A communications tool for computational models. In *Abstracts of Papers of the American Chemical Society*; American Chemical Society: Washington, DC, USA, 2005; Volume 230, pp. U1371–U1372.

International Journal of
Molecular Sciences

MDPI

rticle

Structure Characterization of Zinc Finger Motif 1 and 2 of GLI1 DNA Binding Region

Mousheng Wu, Nusrat Jahan, Amanda Sharp, Anwar Ullah, Corinne E. Augelli-Szafran, Sixue Zhang * and Rebecca J. Boohaker *

Scientific Platforms, Southern Research, 2000 9th Avenue South, Birmingham, AL 35205, USA; moushengwu@hotmail.com (M.W.); nursatjahanbuet@gmail.com (N.J.); aksharp@vt.edu (A.S.); aullah@southernresearch.org (A.U.); caugelli-szafran@southernresearch.org (C.E.A.-S.)

* Correspondence: szhang@southernresearch.org (S.Z.); rboohaker@southernresearch.org (R.J.B.)

Abstract: As a transcription factor, GLI1 plays an important role in cell cycle regulation, DNA replication, and DNA damage responses. The aberrant activation of GLI1 has been associated with cancers such as glioma, osteosarcoma, and rhabdomyosarcoma. The binding of GLI1 to a specific DNA sequence was achieved by five tandem zinc finger motifs (Zif motifs) on the N-terminal part of the molecule. Here, we reported a novel homodimeric crystal structure of Zif1-2. These two Zif motifs are linearized. Namely, Zif2 does not bend and interact with Zif1 of the same molecule. Instead, Zif1 from one molecule interacts with Zif2 from another molecule. The dimer interface of Zif1-2 is unique and different from the conformation of Zif1-2 from the GLI1-DNA co-crystal structure. The dimeric conformation of Zif motifs could represent the native conformation of apo form GLI1 Zif motifs in the cell. The molecular dynamics simulation result of the homodimer, the in silico mutagenesis, and the predicted protease stability of these mutants using a large language model are also presented.

Keywords: oncology; structural biology; DNA binding protein; GLI1; molecular dynamics simulations; large language model

1. Introduction

GLI1, a transcription effector at the terminal end of Hedgehog signaling pathways, was originally identified as a highly expressed gene in a malignant glioma [1]. The expression of GLI1 is tightly regulated during embryonic development and tissue differentiation. The aberrant activation of GLI1 has been linked to the promotion of oncogenic activities such as metastasis, DNA damage repair, stemness, and chemotherapeutic resistance in various cancers such as glioma, pancreatic cancer, and colorectal cancer. The increased expression of GLI1 in various cancers is an oncogenic biomarker and a significant therapeutic target for the treatment of multiple cancers [2,3].

Human GLI1 (~1106 amino acids) contains a conserved DNA binding region at the N-terminal region, a SUFU-interacting domain, a nuclear export sequence (NES), a nuclear location signal (NLS), and a transactivation domain (TAD) [4–6]. The GLI1 DNA binding region consists of five highly conserved tandem C2H2 zinc finger (Zif) motifs which are responsible for binding to promoters of its downstream target genes [7]. The co-crystal structure of GLI1 Zif motifs with DNA (GLI1-DNA) (PDB ID: 2GLI) was determined [8]. The structure shows that Zif1-3 interacts with the phosphate backbone and contributes to the binding stability and recruitment of co-regulators, while Zif4-5 recognizes the consensus sequence 5′-GACCACCCA-3′ in the promoter regions of target genes, thus regulating gene transcription. In the GLI1-DNA structure, Zif1-2 has less contribution to DNA binding than Zif3-5 [8]. Zif1-2 may open new opportunities for discovering selective inhibitors targeting GLI1. Inhibitors targeting the DNA binding groove of GLI1 (mainly consisting of Zif3-5) may have off-target binding to other DNA binding proteins processing similar

DNA binding grooves. Therefore, Zif1-2 of GLI may be a more selective binding region than Zif3-5 for designing and developing direct GLI inhibitors.

Although the Hedgehog signaling pathways, including GLI1, are promising targets for cancer treatment [9], drug discovery efforts directly targeting GLI1 are relatively limited. Direct GLI1 inhibitors such as GANT61 [10], SRI-38832 [3], and JC19 [11] have been reported. However, so far, none of them are in the advanced preclinical or clinical development phase. Of note, the putative binding region of GANT61, SRI-38832, and JC19 is Zif4-5 of GLI. To aid in the discovery of a new generation of GLI1 inhibitors targeting Zif1-2, we determined the crystal structure of the apo form of the first two Zif motifs of GLI1. Interestingly, Zif1 and Zif2 are in a linear configuration and form a unique homodimer in which Zif1 from one molecule interacts with Zif2 from another molecule, and vice versa. The loop regions are also involved in dimerization. This configuration is referred to as the linearized configuration to distinguish it from the bent configuration of Zif1-2 in a DNA-bound state. The bent configuration refers to a configuration in which Zif1 interacts with Zif2 of the same molecule. The *apo* form of Zif1-2 motifs offers a more distinctive structural insight than the DNA-bound GLI1 structure reported before.

2. Results

2.1. Structure Determination

The corresponding residues in the structure of GLI1 zinc motifs (PDB ID: 2GLI) were re-numbered to that of the protein sequence from P08151 in the UNIPROT database (www.uniprot.org, accessed on 1 November 1988). For example, E103 in the structure was renumbered to E234. The Zif1-2 (residue 234–302) from 2GLI [8] was used as a model to determine the crystal structure of GLI1 (234–302). The initial model from Phaser [12] was refined by refmac5 [13] and the electron density was checked under COOT [14]. To reduce atom clashes in the initial model, we decided to use ARP/wARP to rebuild the Zif1-2 model automatically using the phase from molecular replacement. The rebuilt model underwent geometry refinement and was then determined to be the final conformation model of the GLI1 Zif1-2 dimer crystal. In the final model, the two Zif1-2 monomers adapted a linear conformation. Namely, the two monomers aligned side-by-side with the N-terminal of one monomer facing the C-terminal of the second monomer (Figure 1).

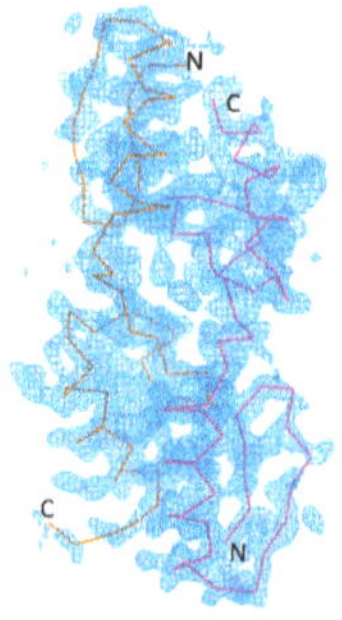

Figure 1. Annealing omit electron density maps with 1.5δ of Zif1-2 models. Pictures were generated using PyMol (The PyMOL Molecular Graphics System, Version 3.0, Schrödinger LLC). Yellow and purple lines indicate the protein backbone of monomer A and B of the dimer crystal structure respectively. Letters N and C mark the beginning of the N-terminus and the end of the C terminus respectively.

2.2. Overall Structure of Zif1-2

There are two molecules of Zif1-2 in the asymmetric unit in the crystal structure of Zif1-2 (Figure 2). Two zinc ions in each Zif molecule are clearly observed in the crystal structure. Both the zinc ions are tetrahedrally coordinated by two histidines (H291 and H295) and two cysteines (C270 and C275). In the first Zif1-2 molecule, the residues 301-302 are missing in the structure. In the other molecule, the first residue, E234, and residues 297-302 are missing in the structure. Both zinc finger motifs are a typical zinc finger structure which contains two helices. The N-terminal starts with a small loop and the C-terminal ends with a long loop (Figure 2). The structures of both Zif motifs are identical to the structure from 2GLI. Interestingly, Zif1 and Zif2 from both molecules in the crystal structure are linearized, while Zif1 and Zif2 are packed against each other in the GLI1-DNA structure. The Zif1 and Zif2 motifs are opened-up in the structure and form an anti-parallel homodimer with the other linearized Zif1-2 molecule. Zif1 from one molecule is packed against the Zif2 from the other molecule, and vice versa, while in the DNA-bound GLI1 structure, Zif1-2 is a monomer instead of a dimer.

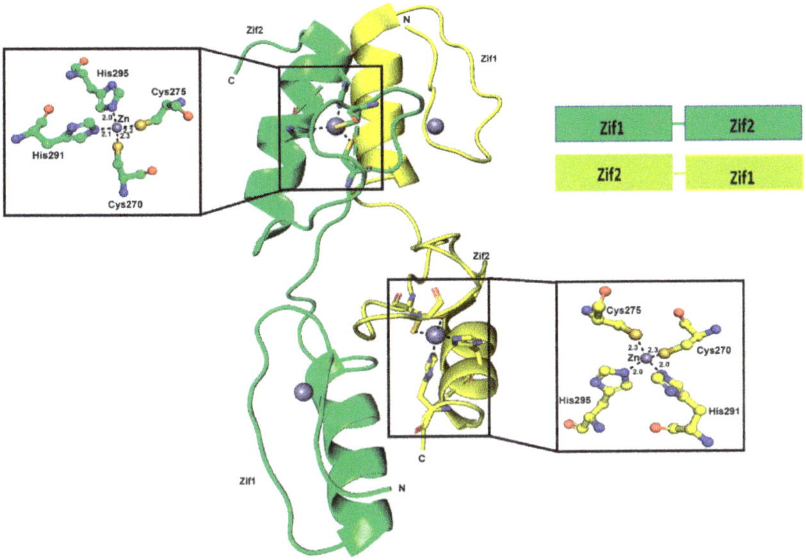

Figure 2. The overall structure of Zif1-2. For simplicity, only two of the four zinc ions are shown in detail. The two monomers are shown in two different colors (green and yellow).

2.3. Dimerization of Zif1-2 and Zif1-5

In the monomer GLI1 Zif1-5 structure (PDB ID: 2GLI) [8], Zif1 and Zif2 are connected by a hinge region G(263)ERK(266). Zif1 is in close contact with Zif2, and Zif2 is involved in DNA binding, while Zif1 is not. Hydrophobic residues from Zif1, W239, L252, V253, I256, and I261, form a hydrophobic patch to interact with another hydrophobic patch on Zif2, which is formed by residues W272, L288, V289, and M292 (Figure 3a). On the other side of the Zif2, residues R277, K283, Y286, R293, and R294 interact with the DNA molecule [8] (Figure 3a). In this Zif1-2 dimer crystal structure, the organization of Zif1 and Zif2 is less bent over compared to the 2GLI structure. Although Zif1 and Zif2 are opened-up, the same extensive hydrophobic interactions still contribute to Zif1-2 dimerization. Two major hydrophobic interaction networks, Zif1-Zif2 and the hinge region interaction networks, contribute to the dimerization interface. The same set of residues from the Zif1 and Zif2 interface described in the 2GLI structure are involved in the dimerization, while the hydrophobic patch of Zif1 interacts with that of Zif2 from the other molecule but not from the same molecule (Figure 3b). In the hinge region, residues of the hinge regions (R265) from both the molecules and residues (E267, F268, V269, H271, L279, and P281) of β-hairpins from each Zif2 motif form an extensive hydrophobic interaction, significantly contributing to the dimerization. Each Zif1-2 molecule contributes an approximate 1279 Å^2 surface area for the dimerization interface. The extensive hydrophobic interaction between the dimers indicates that Zif1-2 can form a stable dimer in the solution. Consistently, the size-exclusion chromatography data (Figure 4a) indicate that the molecular weight of GLI1 Zif1-2 (234–302) was between that of the monomer (~6 kDa) and dimer (~12 kDa). Other different Zif truncations with 3, 4, or 5 Zif motifs, GLI1(234–326), GLI1(234–358), and GLI1(234–388), were also purified for crystallization. Based on size-exclusion chromatography during purification, these Zif truncations also formed a dimer (Figure 4a). GLI1 (234–388, molecular weight ~8 kDa) was eluted in the same elution volume as the standard protein with the molecular weight 12.4 kDa. However, after performing extensive crystallization screenings, we were not able to crystalize other GLI1 Zif truncations besides Zif1-2, implying lower stability of Zif3-5 dimers than Zif1-2 dimers. One possible reason is the lack of hydrophobic patches of Zif3-5 to stabilize their dimer conformation. When we compared the sequences of all five Zif motifs (Figure 4b), the sequence alignment showed that the hydrophobic patches, L(252)VHHI(261) in Zif1 and L(252)VVHM(261), may play a critical role in Zif1-2 dimerization by stabilizing the dimer through intermolecular hydrophobic contact. Such hydrophobic patches are observed in Zif1 and Zif2 but not in the other Zif motifs (Figure 4b). The hydrophobic interface contributing to the dimerization of Zif1-2 mainly consists of I256, I261, W272, and V289 (Figure 5c). More specifically, the Zif1-2 dimer interface involves 7 hydrogen bonds and 117 nonbonded contacts from both monomers (Figure S1). These interactions are contributed by 22 and 20 residues from monomers A and B, respectively. The strength of these dimer interactions was computationally investigated through an in silico mutagenesis study (details in Section 2.5). Of note, there are positive residues (lysine or arginine involved in the recognition of DNA sequences [8]) in other Zif motifs in the location which corresponds to the hydrophobic patches of Zif1 or Zif2.

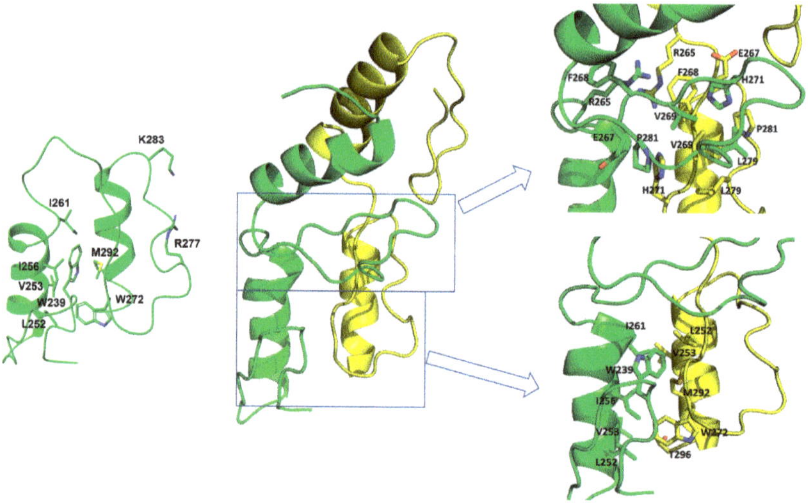

(a) (b)

Figure 3. The interface of Zif1 and Zif2: (**a**) Zif1-2 monomer structure from PDB ID 2GLI. (**b**) The new Zif1-2 dimer structure from PDB ID 7T91. The loop between Zif1 and Zif 2 is highlighted in Magenta. Zif1 is shown in green, and Zif2 in yellow. The interface between Zif1 from molecule A (green) and Zif2 from molecule B (yellow) is shown at the bottom on the right panel. The interface contributed to by the loop region and part of β-hairpin from both molecules is shown on the top on the right panel. All residues involved in the interface are represented by sticks.

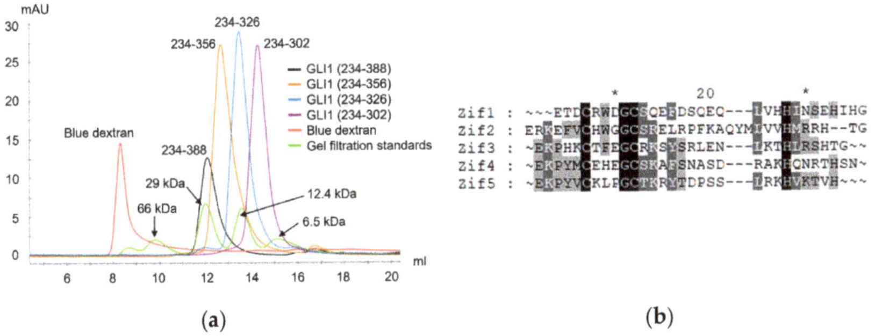

Figure 4. Purification of Zif truncations through size-exclusion chromatography and sequence alignment of all Zif motifs: (**a**) size-exclusion chromatography of all purified Zif truncated proteins; (**b**) sequence alignment of all 5 Zif motifs of GLI1. Sequence alignment was performed using Clustal Omega [15]. Darker color indicates higher sequence consensus with conserved residues among all 5 Zif motifs highlighted in black. Asterisk sign indicates key variant residues.

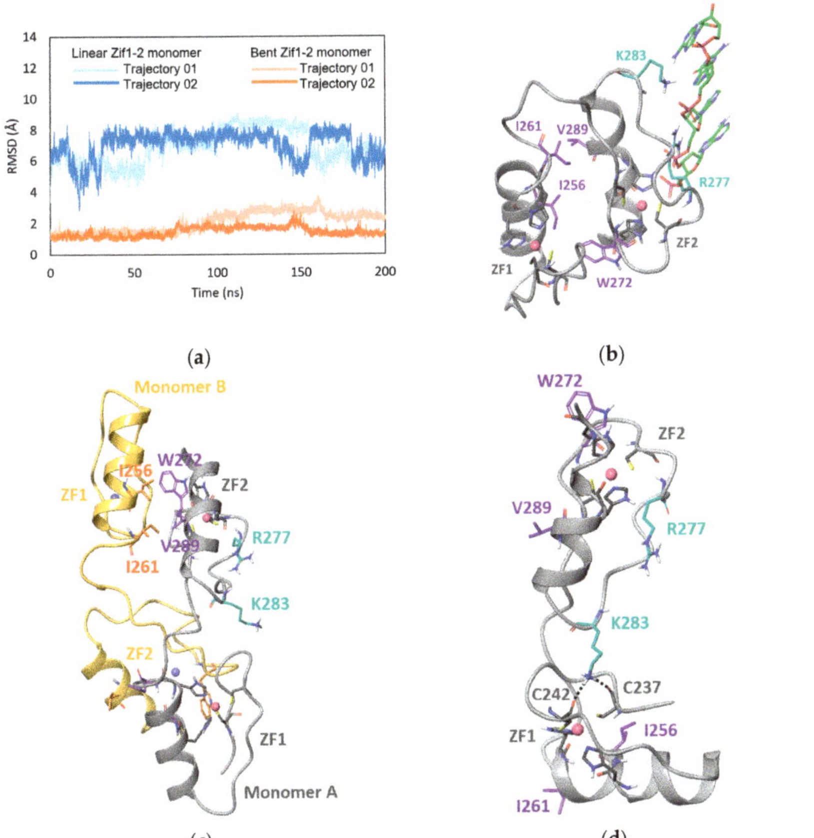

Figure 5. Molecular dynamic simulation studies of Zif1-2: (**a**) RMSD of MD trajectories. (**b**) Bent Zif1-2 from GLI1/DNA crystal complex PDB ID 2GLI. (**c**) Crystal structure of linear Zif1-2 dimer.

Monomer A is indicated by gray ribbon, while monomer B is indicated by yellow ribbon. (**d**) Snapshot at 200ns from MD trajectory 1 of linear Zif1-2. Zinc-chelating residues, hydrophobic interface residues, and DNA binding residues are indicated by gray, purple, and turquoise carbons, respectively. Zinc ions are represented by spheres. DNA is indicated by green carbons (not included in MD simulations). Salt-bridge and hydrogen bond interactions are indicated by dashed black lines. Hydrophobic contacts are indicated by dashed magenta lines. Hydrophobic interface residues from monomer B are indicated by orange carbons.

2.4. MD Simulation Study of GLI1 Zif1-2

MD simulations were performed to further study the two different conformational states of GLI1 Zif1-2: the linear monomer conformation from the apo Zif1-2 dimer crystal structure vs. the bent monomer conformation from the 2GLI/DNA co-crystal structure (PDB ID: 2GLI). The root-mean-square deviation (RMSD) calculated from the MD trajectories was used to evaluate the stability of the two conformational states. After a 100 ns simulation, the linear Zif1-2 monomer entered a state with an average RMSD around 7.3 Å whereas the bent Zif1-2 monomer stayed in a state with an average RMSD around 1.7 Å (Figure 5a). This indicated that the linear Zif1-2 monomer was less stable by itself compared to the bent Zif1-2 monomer. Indeed, an intramolecular hydrophobic interface between Zif1 (I256 and I261) and Zif2 (W272 and V289) was presented in the bent monomer in both the crystal structure and MD trajectories, which was supposed to stabilize the conformation (Figure 5b). In contrast, the crystal structure showed that linear Zif1-2 formed an intermolecular hydrophobic interface with I256 and I261 from Zif1 of monomer B and W272 and V289 from Zif2 of monomer A and vice versa (Figure 5c). Such intermolecular hydrophobic contacts were not presented in the MD trajectory of the linear monomer. In addition, during the 100 ns simulation of the linear monomer, the nonpolar residues I256, I261, W272, and V289 faced the solvent phase, not forming intramolecular hydrophobic contacts with other nonpolar residues either. Thus, the monomer in the linearized conformation did not have such a hydrophobic interface for stabilization. Of note, in the *apo* form, the DNA-binding residues R277 and K283 of the linear monomer were able to form random hydrogen bonds with other residues. One example was the hydrogen bonds of K283 with backbones of C237 and C242 (Figure 5d). Overall, the MD simulations indicates that the linear conformation monomer is not stable by itself, and thus, has the need to form dimer for stabilization.

2.5. In Silico Mutagenesis

Following the MD simulations, the importance of the hydrophobic interface (I256:I261 and W272:V289) was further examined through in silico mutagenesis. The calculated dimer interaction free energy was used as an implication of the interaction strength between the two monomers (Table 1). The more negative value of the interaction free energy implied a stronger interaction between the two monomers. Taking mutation W272G as an example, it had a loss of 35.66 kcal/mol free energy compared to the wild type (−114.26 vs. −149.92 kcal/mol). This implied that, without the nonpolar side chain of W272 to form hydrophobic contact, the dimer interface was less strong. Similarly, W272N, with a polar side chain not able to form hydrophobic contact, also saw a loss of 31.27 kcal/mol free energy; in other words, it was less stable than the wild-type dimer. Likewise, I256G and I261G resulted in a loss of dimer interaction free energy as well. Interestingly, V289G did not result in a loss of free energy. This might imply that the hydrophobic contact centered around W272 contributed more to dimer stability than the hydrophobic contact centered around V289G. Other mutations also resulted in changes in interaction free energy, indicating that the network of hydrophobic interactions collectively drives Zif1-2 dimerization.

Table 1. Impact of mutations on Zif1-2 dimerization interaction free energy.

Mutation	Predicted Interaction Free Energy (kcal/mol)
Wild Type	−149.92
W239G	−101.03
W239N	−123.23
L252G	−140.63
L252S	−142.30
V253G	−143.68
V253S	−137.70
I256G	−145.48
I256S	−142.37
I261G	−140.80
I261N	−146.78
I261S	−142.44
W272G	−114.26
W272N	−118.65
W272Y	−142.84
V289G	−151.93
V289T	−160.63
M292G	−133.47
M292N	−135.87
T296S	−147.12

In order to see whether the above mutations of interest would be stable enough to be produced and studied in a wet lab, we leveraged the large language model Protein-BERT [16] to predict the protease susceptibility of all possible mutations (Figure 6). The predictions indicated that the mutations of interest, especially W272G and W272N, were stable mutations. These two mutations were predicted to be even more stable than the wild-type GLI. This may be due to the removal of a protease-recognizable tryptophan residue. It would be warranted in the future to experimentally determine whether W272G and W272N indeed disrupt Zif1-2 dimerization as hypothesized.

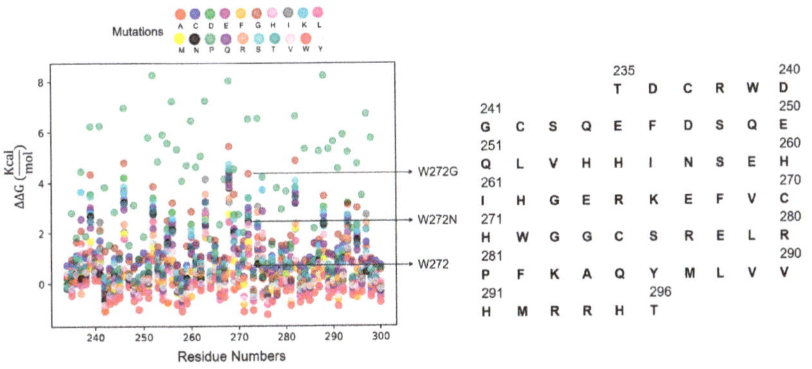

Figure 6. Stability prediction of Zif1-2 mutants using the large language model. The wild-type GLI sequence used for this prediction is listed on the right.

3. Discussion

In the structure of GLI1-DNA in which Zif1-5 is a monomer, Zif1 is in close contact with Zif2 (Figure 3a), adopting a bent conformation. When Zif1-2 from 2GLI is superimposed onto our Zif1-2 dimeric structure, the RMSD of the main chain (residues 234–262) is only 0.855 Å between Zif1 from 2GLI and Zif1 of molecule A in our Zif1-2 structure. However, Zif2 (residues 268–296) from 2GLI is located on Zif2 of molecule B (Figure 7a). From the structure comparison (Figure 7b), the Zif2 from 2GLI needs to rotate about 180 degrees to form a linearized Zif1-2, as seen in the dimeric Zif1-2 structure. The binding of DNA may restrain Zif1-2 from forming a linearized dimer. The Zif1-2 dimer crystal was determined in the absence of DNA. Moreover, the fact that the Zif1-2 apo dimer was observed in size-exclusion chromatography (Figure 4a) indicates that such a dimer is likely to be stable in both the solution and crystal phases.

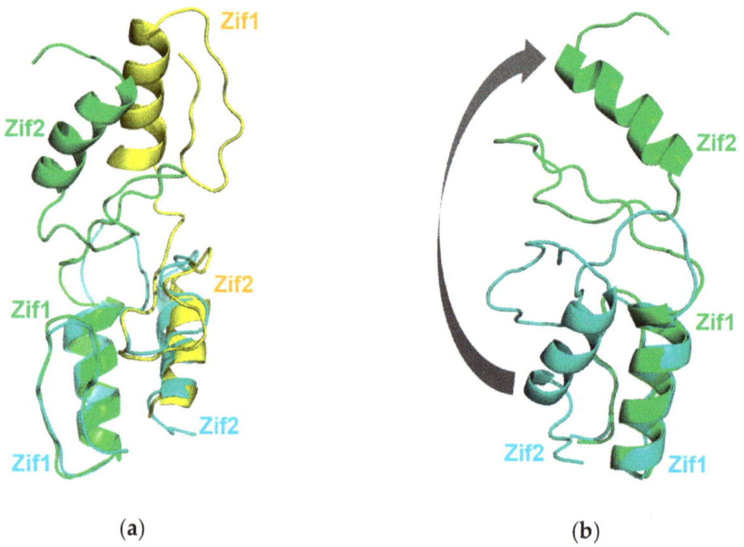

Figure 7. A comparison of the dimeric Zif1-2 structure with the Zif1-2 structure from the 2GLI structure: (**a**) superimposition of Zif1-2 from the 2GLI structure with the dimeric Zif1-2 structure; (**b**) superimposition of Zif1-2 from the 2GLI structure with molecule A of the dimeric Zif1-2 structure. Zif1-2 from the 2GLI structure is shown in cyan. The two molecules of the dimeric Zif1-2 structure are shown in green and yellow, respectively. The arrow shows the moving direction of Zif2 when Zif1-2 forms a linear Zif1-2 conformation.

4. Materials and Methods

4.1. Molecular Cloning, Protein Expression, and Purification

The yeast SUMO family protein SMT3 gene followed by the TEV cleavage site (SUMO-TEV) was amplified from a pET-His6-SUMO-TEV-LIC plasmid (Addgene #: 29659) via PCR using Phusion high-fidelity DNA polymerase (Thermo Fisher Scientific, Waltham, MA, USA), a forward primer (GGAATTCCATATGTCGGACTCAGAAGTCAATCAAGAAGC), and a reverse primer (CGCGGATCCGGATTGGAAGTACAGGTTTTCCTCGATC). The NdeI restriction site in the forward primer and BamHI site in the reverse primer were underlined. The PCR product was purified using a GeneJET gel extraction kit (Thermo Fisher Scientific) and digested by NdeI and BamHI restriction enzymes. The digested product was purified using a GeneJET PCR purification kit (Thermo Fisher Scientific) and ligated into a pET28a(+) vector, which was digested by NdeI and BamHI enzymes, using T4 DNA ligase (Thermo Fisher Scientific). The ligated product was transformed into NovaBlue *E. coli* competent cells (Novagen), and the plasmids from single colonies were purified by using a GeneJET plasmid miniprep kit (Thermo Fisher Scientific). The insertion of SUMO-TEV was confirmed by DNA sequencing. The plasmid was named the pET28a-SUMO-TEV vector. Human GLI1 truncation containing zinc finger motif 1 and 2 (residue 234–302) were amplified from a longer human GLI1 zinc finger motif construct (residue 232–391) in a pHis-parallel1 vector in our laboratory using the primers in Table 1. The PCR product was digested by BamHI and SalI restriction enzymes and purified as described above. The digested product was ligated to a pET28a-SUMO-TEV vector. The plasmid was prepared from NovaBlue *E. coli* cells (Novagen, Madison, WI, USA) and transformed into *E. coli* Rosetta pLysS competent cells (Novagen) for protein expression. A single colony was inoculated in Luria–Bertani (LB) broth medium and cultured at 37 °C overnight. A 10 mL overnight culture was added to 500 mL of auto-induction medium [9], and the cells were cultured at 37 °C for approximately five hours. The temperature was lowered to 18 °C and the cells were further cultured overnight. The cells were harvested and stocked at −80 °C.

For protein purification, the frozen cell pellet was thawed and resuspended in lysis buffer (20 mM HEPES, pH 7.6, 500 mM NaCl, 0.1 mM TCEP) with 20 mM imidazole, a tablet of EDTA-free protease inhibitors (Thermo Fisher Scientific) and 1 mg/mL lysozyme. The cells were disrupted by sonication, and the cell pellets were spun down at 9500 rpm for 1 h. The supernatant was loaded onto a Ni-NTA column (Cytiva, Marlborough, MA, USA). The Ni-NTA column was then washed stepwise in lysis buffer with 20 mM, 40 mM, and 60 mM imidazole. His-SUMO-GLI (234–302) protein was eluted in lysis buffer with 400 mM imidazole. An amount of 10 mg GST-fused TEV protease solution prepared in the lab was added to the GLI1 elution, and the protein solution was passed through a G-25 desalting column (Cytiva). The protein solution with TEV protease was incubated at 4 °C overnight and then reloaded onto a glutathione column (Cytiva) to remove GST-TEV. The protein solution was diluted in 20 mM Tris pH 8.0 buffer to reach 350 mM NaCl. The GLI1 was further purified by a HiTrap Q HP column (Cytiva) using Buffer A (20 mM Tris, 300 mM NaCl, 0.1 mM TCEP, pH 8.0) and Buffer B (20 mM Tris, 1 M NaCl, 0.1 mM TCEP, pH 8.0). The fractions containing GLI1(234-302) were collected and passed through a Superdex 75 column (Cytiva) in size-exclusion buffer (20 mM HEPES, pH 7.6, 500 mM NaCl, 0.1 mM TCEP). The protein quality from all purification steps was analyzed through SDS-PAGE.

Other Zif truncations, GLI1(234–326), GLI1(234–358), and GLI1(234–388), were cloned using the molecular cloning protocol described above using the forward primer of GLI1(234–302) (Table 2). The reversed primers for GLI1(234–326), GLI1(234–358), and GLI1(234–388) were as follows: GLI1_326SalI (CGCGTCGACTTACGTGTGTGACCGCAGGTGCGTCTTC), GLI1_358SalI (CGCGTCGACTTAATTGGAATGGGTCCGATTCTGGTGC), and GLI1_388SalI (CGCGTCGACTTAACCATGCACTGTCTTGACATGTTTTCG), respectively. All GLI1 Zif truncated proteins were purified using the same protocol as for GLI1(234–302).

Table 2. Macromolecule production information of GLI1(234–302).

Parameter	Value
Source organism	*Homo Sapiens*
DNA source	Laboratory plasmid
Forward primer	CGCGGATCCGAAACTGACTGCCGTTGGGATGGC
Reverse primer	CGCGTCGACTTACTTGTGTGGCTTCTCGCCAGTGTG
Cloning vector	pET28a (+)-SUMO-TEV
Expression vector	pET28a (+)-SUMO-TEV
Expression host	*E. coli* Rosetta pLysS
Sequence of full construct	MGSSHHHHHHSSGLVPRGSHMSDSEVNQEAK PEVKPEVKPETHINLKVSDGSSEIFFKIK KTTPLRRLMEAFAKRQGKEMDSLRFLYDGIRIQADQT PEDLDMEDNDIIEAHREQIGGIEENLYFQSGS ETDCRWDGCSQEFDSQEQLVHHINSEHIH GERKEFVCHWGGCSRELRPFKAQYMLVVHMRRHTGEKPHK

4.2. Crystallization

Fractions containing the GLI1(234–302) protein from size-exclusion chromatography were collected and concentrated to 5.4 mg/mL using an Amicon 3 kDa molecular-weight cutoff centrifugal concentrator. The concentrated protein was used to set up crystallization screenings using a Phoenix (ArtRobbins Instrument, Sunnyvale, CA, USA) crystallization robot in a 96-well sitting-drop Intelli-plate (ArtRobbins Instrument, Sunnyvale, CA, USA) by mixing 200 nL protein with 200 nL crystallization solutions. The crystallization screen kits in 96-deep wells used for the screening included Crystal Screen, Index, MembFrac, Natrix, PEG/ION, PEGRx, and SaltRx (Hampton Research, Aliso Viejo, CA, USA). The crystallization plates were incubated in a crystallization chamber at 17 °C for at least 3 days. The crystallization plates were examined under a microscope. The initial crystals were grown in Index condition D07 (25% PEG3350, 0.1M Bis-Tris pH 6.5). The crystallization condition was optimized to 29–32% PEG3350, 0.1M Bis-Tris pH 6.5 in a 24-well sitting-drop plate (Hampton Research) under 1 µL protein with 1 µL reservoir solution drop. The crystallization information is summarized in Table 3.

Table 3. Crystallization of GLI1(234–302).

Parameter	Value
Method	Sitting-drop
Plate type	24-well plate
Temperature (K)	291
Protein concentration (mg/mL)	5.4
Buffer for protein solution	20 mM HEPES, pH = 7.6, 500 mM NaCl, 0.1 mM TCEP
Reservoir solution	29–32% PEG3350, 0.1 M Bis-Tris pH = 6.5
Volume (µL) and ratio of drop	1:1
Volume of reservoir (µL)	500

4.3. *Data Collection and Processing*

GLI1(234–302) crystals were cryo-protected in 30% PEG3350, 0.1 M Bis-Tris pH 6.5, and 20% glycerol and frozen in liquid nitrogen. X-ray diffraction data were collected at SER-CAT (Southeast Regional Collaborative Access Team) beamline 22-ID at Advanced Proton Sources (Argonne National Laboratory, Lemont, IL, USA). The data were processed using XDS [17] and scaled using Scala [18]. The data were processed to 2.05 Å and the crystal belonged to space group P65 with the following unit-cell parameters: a = b = 66.72 Å, c = 65.90 Å. The statistics of the data collection and processing are summarized in Table 4.

Table 4. Data processing and refinement statistics.

Section	Parameter	Value *
Data collection	Beamline	22-ID, APS
	Processing software	XDS version 2019
	Wavelength (Å)	1.0
	Resolution range (Å)	33.36–2.05 (2.123–2.05)
	Space group	P65
	Unit cell a, b, c (Å)	66.728, 66.728, 65.909
	Unit cell α, β, γ (°)	90, 90, 120
	Total reflections	21,071 (2088)
	Unique reflections	10,536 (1042)
	Multiplicity	2.0 (2.0)
	Completeness (%)	99.88 (100.00)
	Mean *I/sigma* (*I*)	17.64 (4.41)
	Wilson B factor (Å^2)	40.82
	R-merge	0.03403 (0.1799)
	R-means	0.04813 (0.2544)
	R-pim	0.03403 (0.1799)
	$CC_{1/2}$	0.998 (0.899)

Table 4. *Cont.*

Section	Parameter	Value *
Refinement	Reflections used in refinement	10,527 (1044)
	Reflections used for R_{free}	509 (56)
	R_{work}	0.1985 (0.2403)
	R_{free}	0.2258 (0.2982)
	CC_{work}	0.948 (0.785)
	CC_{free}	0.934 (0.700)
	Number of nonhydrogen atoms	1160
	Macromolecules	1084
	Ligands	4
	Solvent	72
	Protein residues	129
	Bond length (Å)	0.013
	Bond angle (°)	1.60
Ramachandran plot	Ramachandran favored (%)	99.20
	Ramachandran allowed (%)	0.80
	Ramachandran outliers (%)	0.00
	Rotamer outliers (%)	1.69
	Clash score	3.85
	Average B factor (Å^2)	48.57
	Macromolecules	48.50
	Ligands	39.66
	Solvent	50.03
	Number of TLS groups	2
	PDB ID	7T91

* Values in parentheses are for the highest-resolution shell.

4.4. Structure Solution and Refinement

Initially, the structure of zinc finger motifs 1 and 2 (Zif1-2) without zinc ions (residue 234–302) from the structure of the GLI1 zinc finger motifs (PDB ID: 2GLI) was used as a search model to determine the crystal structure of GLI1(234–302) using Phaser-MR [12]. Two molecules were found during the model search with LLG = 57 and LLG = 146. The initial model then was refined by Refmac5 [13]. The refinement was stuck with a very high Rfree and R factor, 44.78 and 46.92, respectively. Then, the phase from molecular replacement was used to rebuild the model of Zif1-2 using ARP/wARP [19] with the function of automated model building starting from the experimental phases. The R factor of the final ARP/wARP model was 27.79. The model was then refined using Refmac5 to an R factor of 26.38, and an Rfree factor of 28.96. The model was checked and modified in COOT [14]. After several rounds of refinement by Refmac5 and model building in COOT, zinc ions and water molecules were added into the model. The final model was refined to an R factor of 19.8 and an Rfree factor of 22.5 with 99.2% most favored residues and 0.8% allowed residues in the final model checked by Procheck [20]. The statistics of the model refinement are summarized in Table 4. The solved structure was deposited in the PDB Data Bank (www.rcsb.org) with PDB ID: 7T91.

4.5. MD Simulation

The bent Zif1-2 monomer used for MD simulations was extracted from the GLI1/DNA co-crystal (PDB ID: 2GLI) and was prepared using the Protein Preparation Wizard [21] in Schrödinger Life Science Suite (Release 2020-1, Schrödinger, LLC: New York, NY, USA). Hydrogen atoms were added to the structure. The PROPKA module [22] in Schrödinger was used to assign the protonation states of the protein residues at pH = 7.4. The linear Zif1-2 monomer was extracted from the *apo* Zif1-2 dimer crystal structure (PDB ID: 7T91) and was prepared for MD simulations using the same procedure as above. MD simulations were performed using the GPU-accelerated AMBER Molecular Dynamics Package [23,24] following the protocol described in the supplementary information. Briefly, an orthogonal solvent box of a TIP3P water model was added to the protein system with a 15 Å thickness on each side. One sodium ion was added to the solvent to neutralize the charge. The system was treated with AMBER force field 12 and a periodic boundary condition. The zinc-chelating motifs were treated with the ZAFF parameters [25,26]. The system underwent seven-step minimization first and then was gradually heated up to a temperature of 300 K under restraints. Subsequently, 10 ns restraint-free equilibration was performed using the NVT ensemble. Once equilibration was reached, a 200 ns MD simulation was carried out using the NVT ensemble. For each system (bent and linear Zif1-Zif2), two independent 200 ns MD trajectories were produced. The temperature and energy of the systems remained stable during the 200 ns simulations (Figure S2). After the MD simulations, the CPPTRJ tool [27] in the AMBER package was used to perform the root-mean-square fluctuation analysis. The crystal structures of bent and linear Zif1-2 were used as the reference frame when calculating the RMSD. The Cα carbon atoms of the protein backbone were used to calculate the RMSD of each frame to the reference frame.

4.6. In Silico Mutagenesis and Free Energy Calculation

The Schrödinger Small Molecule Drug Discovery Suite was used to perform the in silico mutagenesis and free energy calculations. After each residue was mutated, molecular mechanics energy minimization was performed on the mutated residue via the Prime module [28] of Schrödinger to optimize its new conformation. Subsequently, the free-energy calculations were performed using the MM-GBSA method implemented in the Prime module.

4.7. Large Language AI Model to Predict Protein Stability

To predict the stability upon a single mutation in the Zif1-2 sequence, we used ProteinBERT [16], an open-source LLM that has been pretrained (~106 M known protein sequences to learn protein representation) and fine-tuned (~15 K protein sequences with protease susceptibility assay data) for stability predictions. ProteinBERT exhaustively assessed the impact of every type of amino acid mutation at each position of the given Zif1-2 sequence. There were 1341 single-site mutants for Zif1-2 with a length of 66 AAs (residue numbers 234 to 300 from PDB ID 7T91). The open-source code to run ProteinBERT was obtained from the code repository Github: https://github.com/nadavbra/protein_bert (accessed on 17 December 2023).

5. Conclusions

A dimeric Zif1-2 crystal structure is reported here. Zif1-2 forms a unique dimer interface which is totally different from that of the 2GLI structure, though they both use a similar hydrophobic interaction network. The dimer conformation may also be presented in the whole Zif motif (Zif1-5) purified without denaturation and a refolding process, which uses the same Zif1-2 dimer interface. Our MD simulation study showed that the binding of DNA may prevent Zif1-2 dimer formation. Whether the full-length GLI1 molecule in the cell is dimeric warrants further investigation in the future.

Supplementary Materials: The following supporting information can be downloaded at: https://www.mdpi.com/article/10.3390/ijms252413368/s1.

Author Contributions: Conceptualization, M.W., R.J.B. and S.Z.; methodology, M.W. and S.Z.; software, M.W., N.J., A.S., A.U. and S.Z.; validation, A.U.; formal analysis, M.W., N.J., A.S., A.U. and S.Z.; investigation, M.W. and S.Z.; resources, M.W., R.J.B., C.E.A.-S. and S.Z.; data curation, M.W.; writing—original draft preparation, M.W. and S.Z.; writing—review and editing, M.W., N.J., A.U., C.E.A.-S., S.Z. and R.J.B.; visualization, M.W., N.J. and S.Z.; supervision, R.J.B., C.E.A.-S. and S.Z.; funding acquisition, R.J.B. All authors have read and agreed to the published version of the manuscript.

Funding: This research was funded by THE NATIONAL INSTITUTES OF HEALTH, grant number 5RO1CA183921-05.

Data Availability Statement: The data on the crystal structure are openly accessible at the Protein Data Bank with the entry ID 7T91. The link to the data entry is https://www.rcsb.org/structure/7T91 (accessed on 21 December 2022).

Acknowledgments: We would like to thank the scientists at SER-CAT beamline 22-ID at the Advance Proton Source in the Argonne National Laboratory for their assistance in the X-ray data collection.

Conflicts of Interest: The authors declare no conflicts of interest. The funders had no role in the design of the study; in the collection, analyses, or interpretation of the data; in the writing of the manuscript; or in the decision to publish the results.

References

1. Wong, A.J.; Bigner, S.H.; Bigner, D.D.; Kinzler, K.W.; Hamilton, S.R.; Vogelstein, B. Increased expression of the epidermal growth factor receptor gene in malignant gliomas is invariably associated with gene amplification. *Proc. Natl. Acad. Sci. USA* **1987**, *84*, 6899–6903. [CrossRef]
2. Avery, J.T.; Zhang, R.; Boohaker, R.J. GLI1: A therapeutic target for cancer. *Front. Oncol.* **2021**, *11*, 673154. [CrossRef]
3. Zhang, R.; Ma, J.; Avery, J.T.; Sambandam, V.; Nguyen, T.H.; Xu, B.; Suto, M.J.; Boohaker, R.J. GLI1 inhibitor SRI-38832 attenuates chemotherapeutic resistance by downregulating NBS1 transcription in BRAF(V600E) colorectal cancer. *Front. Oncol.* **2020**, *10*, 241. [CrossRef]
4. Bauer, N.C.; Doetsch, P.W.; Corbett, A.H. Mechanisms regulating protein localization. *Traffic* **2015**, *16*, 1039–1061. [CrossRef]
5. Han, Y.; Shi, Q.; Jiang, J. Multisite interaction with Sufu regulates Ci/Gli activity through distinct mechanisms in Hh signal transduction. *Proc. Natl. Acad. Sci. USA* **2015**, *112*, 6383–6388. [CrossRef] [PubMed]
6. Mastrangelo, E.; Milani, M. Role and inhibition of GLI1 protein in cancer. *Lung Cancer* **2018**, *9*, 35–43. [CrossRef]
7. Ruppert, J.M.; Kinzler, K.W.; Wong, A.J.; Bigner, S.H.; Kao, F.T.; Law, M.L.; Seuanez, H.N.; O'Brien, S.J.; Vogelstein, B. The GLI-Kruppel family of human genes. *Mol. Cell. Biol.* **1988**, *8*, 3104–3113. [CrossRef]
8. Pavletich, N.P.; Pabo, C.O. Crystal structure of a five-finger GLI-DNA complex: New perspectives on zinc fingers. *Science* **1993**, *261*, 1701–1707. [CrossRef] [PubMed]
9. Xie, H.; Paradise, B.D.; Ma, W.W.; Fernandez-Zapico, M.E. Recent advances in the clinical targeting of Hedgehog/GLI signaling in cancer. *Cells* **2019**, *8*, 394. [CrossRef]
10. Zhang, R.; Wu, J.; Ferrandon, S.; Glowacki, K.J.; Houghton, J.A. Targeting GLI by GANT61 involves mechanisms dependent on inhibition of both transcription and DNA licensing. *Oncotarget* **2016**, *7*, 80190–80207. [CrossRef] [PubMed]
11. Maresca, L.; Crivaro, E.; Migliorini, F.; Anichini, G.; Giammona, A.; Pepe, S.; Poggialini, F.; Vagaggini, C.; Giannini, G.; Sestini, S.; et al. Targeting GLI1 and GLI2 with small molecule inhibitors to suppress GLI-dependent transcription and tumor growth. *Pharmacol. Res.* **2023**, *195*, 106858. [CrossRef]
12. McCoy, A.J.; Grosse-Kunstleve, R.W.; Adams, P.D.; Winn, M.D.; Storoni, L.C.; Read, R.J. Phaser crystallographic software. *J. Appl. Crystallogr.* **2007**, *40*, 658–674. [CrossRef]
13. Murshudov, G.N.; Vagin, A.A.; Dodson, E.J. Refinement of macromolecular structures by the maximum-likelihood method. *Acta Crystallogr. D Biol. Crystallogr.* **1997**, *53*, 240–255. [CrossRef] [PubMed]
14. Emsley, P.; Lohkamp, B.; Scott, W.G.; Cowtan, K. Features and development of Coot. *Acta Crystallogr. D Biol. Crystallogr.* **2010**, *66*, 486–501. [CrossRef] [PubMed]
15. Madeira, F.; Madhusoodanan, N.; Lee, J.; Eusebi, A.; Niewielska, A.; Tivey, A.R.N.; Lopez, R.; Butcher, S. The EMBL-EBI job dispatcher sequence analysis tools framework in 2024. *Nucleic Acids Res.* **2024**, *52*, W521–W525. [CrossRef]
16. Brandes, N.; Ofer, D.; Peleg, Y.; Rappoport, N.; Linial, M. ProteinBERT: A universal deep-learning model of protein sequence and function. *Bioinformatics* **2022**, *38*, 2102–2110. [CrossRef] [PubMed]
17. Kabsch, W. Xds. *Acta Crystallogr. D Biol. Crystallogr.* **2010**, *66*, 125–132. [CrossRef]
18. Evans, P. Scaling and assessment of data quality. *Acta Crystallogr. D Biol. Crystallogr.* **2006**, *62*, 72–82. [CrossRef]
19. Langer, G.; Cohen, S.X.; Lamzin, V.S.; Perrakis, A. Automated macromolecular model building for X-ray crystallography using ARP/wARP version 7. *Nat. Protoc.* **2008**, *3*, 1171–1179. [CrossRef] [PubMed]
20. Laskowski, R.A.; MacArthur, M.W.; Moss, D.S.; Thornton, J.M. PROCHECK: A program to check the stereochemical quality of protein structures. *J. Appl. Crystallogr.* **1993**, *26*, 283–291. [CrossRef]
21. Sastry, G.M.; Adzhigirey, M.; Day, T.; Annabhimoju, R.; Sherman, W. Protein and ligand preparation: Parameters, protocols, and influence on virtual screening enrichments. *J. Comput. Aided Mol. Des.* **2013**, *27*, 221–234. [CrossRef] [PubMed]
22. Olsson, M.H.; Sondergaard, C.R.; Rostkowski, M.; Jensen, J.H. PROPKA3: Consistent treatment of internal and surface residues in empirical pKa predictions. *J. Chem. Theory Comput.* **2011**, *7*, 525–537. [CrossRef]
23. Gotz, A.W.; Williamson, M.J.; Xu, D.; Poole, D.; Le Grand, S.; Walker, R.C. Routine microsecond molecular dynamics simulations with AMBER on GPUs. 1. Generalized born. *J. Chem. Theory Comput.* **2012**, *8*, 1542–1555. [CrossRef]
24. Salomon-Ferrer, R.; Gotz, A.W.; Poole, D.; Le Grand, S.; Walker, R.C. Routine microsecond molecular dynamics simulations with AMBER on GPUs. 2. explicit solvent particle mesh Ewald. *J. Chem. Theory Comput.* **2013**, *9*, 3878–3888. [CrossRef]
25. Li, P.; Roberts, B.P.; Chakravorty, D.K.; Merz, K.M., Jr. Rational design of particle mesh Ewald compatible Lennard-Jones parameters for +2 metal cations in explicit solvent. *J. Chem. Theory Comput.* **2013**, *9*, 2733–2748. [CrossRef] [PubMed]

26. Peters, M.B.; Yang, Y.; Wang, B.; Fusti-Molnar, L.; Weaver, M.N.; Merz, K.M., Jr. Structural survey of zinc containing proteins and the development of the zinc AMBER force field (ZAFF). *J. Chem. Theory Comput.* **2010**, *6*, 2935–2947. [CrossRef]
27. Roe, D.R.; Cheatham, T.E., 3rd. PTRAJ and CPPTRAJ: Software for processing and analysis of molecular dynamics trajectory data. *J. Chem. Theory Comput.* **2013**, *9*, 3084–3095. [CrossRef]
28. Jacobson, M.P.; Pincus, D.L.; Rapp, C.S.; Day, T.J.; Honig, B.; Shaw, D.E.; Friesner, R.A. A hierarchical approach to all-atom protein loop prediction. *Proteins* **2004**, *55*, 351–367. [CrossRef]

International Journal of
Molecular Sciences

MDPI

rticle

Comparative Analysis of Symmetry Parameters in the E2 Inner Core of the Pyruvate Dehydrogenase Complex

an-ul Kim [1,2,†], Myeong Seon Jeong [1,3,†], Mi Young An [1], Yoon Ho Park [1], Sun Hee Park [4,5], Sang J. Chung [4,5], oon-sun Yi [3], Sangmi Jun [3], Young Kwan Kim [2,*] and Hyun Suk Jung [1,2,*]

1 Department of Biochemistry, College of Natural Sciences, Kangwon National University, Chuncheon 24341, Republic of Korea
2 Kangwon Center for Systems Imaging, Chuncheon 24341, Republic of Korea
3 Center for Bio-Imaging Translational Research, Korea Basic Science Institute, Cheongju 28119, Republic of Korea
4 School of Pharmacy, Sungkyunkwan University, Suwon 16419, Republic of Korea
5 AbTis Co., Ltd., Suwon 16648, Republic of Korea
* Correspondence: youngk@kangwon.ac.kr (Y.K.K.); hsjung@kangwon.ac.kr (H.S.J.)
† These authors contributed equally to this work.

Abstract: Recent advances in cryo-electron microscopy (cryo-EM) have facilitated the high-resolution structural determination of macromolecular complexes in their native states, providing valuable insights into their dynamic behaviors. However, insufficient understanding or experience with the cryo-EM image processing parameters can result in the loss of biological meaning. In this paper, we investigate the dihydrolipoyl acetyltransferase (E2) inner core complex of the pyruvate dehydrogenase complex (PDC) and reconstruct the 3D maps using five different symmetry parameters. The results demonstrate that the reconstructions yield structurally identical 3D models even at a near-atomic structure. This finding underscores a crucial message for researchers engaging in single-particle analysis (SPA) with relatively user-friendly and convenient image processing software. This approach helps reduce the risk of missing critical biological details, such as the dynamic properties of macromolecules.

Keywords: cryo-electron microscopy; single-particle analysis; structural dynamics; macromolecule; pyruvate dehydrogenase complex

1. Introduction

Cryo-electron microscopy (cryo-EM) has made significant contributions to the field of structural biology, particularly since nearly a decade ago during the so-called "resolution revolution" [1,2]. This transformative period was marked by the advent of direct electron detectors (DEDs), high-end transmission electron microscopy (TEM), and, simultaneously, the development of software packages for image processing [3,4]. These advancements collectively propelled cryo-EM into becoming a groundbreaking method in structural biology, allowing for unprecedented insights into molecular structures and biological phenomena [5,6]. Although traditional methods in structural biology have made significant contributions over the decades, they offer limited abilities to conduct comprehensive structural determinations of relatively high-molecular-weight (MW) proteins, protein complexes, particularly large macromolecules, and viruses [7]. Consequently, methods that complement the shortcomings of established techniques, such as X-ray crystallography and nuclear magnetic resonance (NMR), have received considerable attention [8].

From the perspective of structural biology, which includes the study of structural dynamics, analyzing the structures of proteins in their native state is crucial [9,10]. Moreover, a comprehensive structural analysis of the target in its native complex state can provide valuable information about the inherent protein–protein interactions (PPIs) of the protein or protein complex [11]. Understanding PPIs is crucial to grasp both their functionality and stability [12]. Additionally, an understanding of these factors can provide important insights not only into the protein structures but also their own structural specification [13]. In this regard, the most widely used method for the high-resolution structural analysis of proteins using cryo-EM is single-particle analysis (SPA) [14]. SPA is a powerful approach for obtaining near-atomic-resolution structural information about proteins or macromolecules [15,16]. Owing to the numerous image processing software packages developed over the past few decades, SPA has become relatively straightforward to use to deliver rapid and excellent results [17,18]. However, given the enhanced accessibility and user-friendliness of image processing procedures for SPA, users should strive to exclude subjective opinions and maximize the utilization of structural information to achieve the actual reconstruction of protein structures [19,20].

With the aim of proposing an approach to improve the processing results, we focused on the difference among the various symmetry parameters in SPA to analyze the dihydrolipoyl acetyltransferase (E2) inner core complex of the pyruvate dehydrogenase complex (PDC) and verified the results based on these symmetry parameters. We study the influence of different symmetry parameters on the 3D reconstruction of the E2 complex by using five parameters: asymmetry (C1), cyclic5 (C5), dihedral2 (D2), tetrahedral (T), and icosahedral (I). These findings convey an important message about the details users might overlook in the 3D reconstruction of a dynamic protein structure using SPA with cryo-EM. Ultimately, the accurate application of symmetry parameters is essential not only for obtaining better maps but also for enhancing our understanding of structural dynamics, thereby facilitating a deeper analysis of the biological relevance of protein assemblies.

2. Results

2.1. Structural Features of the E2 Complex of *B. stearothermophilus*

The E2 protein of *B. stearothermophilus* is a protein with an MW of approximately 43 kDa. It is a highly conserved protein comprising the lipoyl-binding domain (LD, 2-77) located in the N-terminal region, the peripheral subunit-binding domain (PSBD, 130-167) situated in the middle region, and the catalytic acetyltransferase core domain (CD, 214-425) positioned in the C-terminal region (Figure 1A) [21]. To enhance the yield of protein expression and complex formation through *E. coli*, we constructed a region containing only the CD by including amino acid 185 to 425, but excluding the LD and PSBD [22,23]. As shown in Figure 1B, we confirmed the monomeric form at approximately 32 kDa on SDS-PAGE. The purified truncated E2 was observed to appropriately form complexes in the negatively stained microscopic field; we observed a structure similar to that of several previously published E2 complexes, as depicted in Figure 1C. Additionally, through unstained frozen hydrated micrographs for 3D reconstruction of the E2 complex using cryo-EM, we confirmed a size of approximately 25 nm. As a result, by utilizing 2D classification, we observed the E2 complex in high resolution from various angles, as illustrated in Figure 1D.

2.2. Reconstruction of the E2 Complex, Determined Using Various Applied Symmetry Parameters

The 3D reconstruction of E2 was performed to investigate the impact of the symmetry parameters on the image processing results. For this purpose, we used 2D classification to select a final particle set consisting of 69,946 particles. Using the model generated by the same ab initio reconstruction as the initial model, we carried out the process to determine the values of five different parameters: C1, C5, D2, T, and I. The final version of the E2 cryo-EM model achieved the highest resolution through reconstruction using the I parameter, and each reconstruction yielded the following resolution values: C1 = 4.93 Å, C5 = 4.76 Å, D2 = 4.79 Å, T = 4.36 Å, and I = 4.20 Å (Figures 2 and 3). The units constituting the entire complex were consistently reconstructed identically in all five models. Many previous studies on the structural analysis of various E2 complexes predominantly utilized the symmetric restraint I for reconstruction [22,24,25]. However, our results demonstrated that not only this symmetric restraint, which is commonly considered, but also parameters such as C1, C5, D2, and T are similar to the previous results of the E2 complex. The comparative results in Figure 4A, superposed based on C1 as the center, are displayed to illustrate these findings. Due to slight differences in the resolution, it is not possible to discern the differences in perfect detail. However, it can be acknowledged to a satisfactory extent that the five maps were reconstructed into the same 3D shapes. Moreover, for a more detailed comparison, the E2 trimeric complex, which is composed of three E2 monomers, was segregated to ascertain that, in combination, their 3D structures in fact constituted that of the entire trimeric complex (Figure 4B).

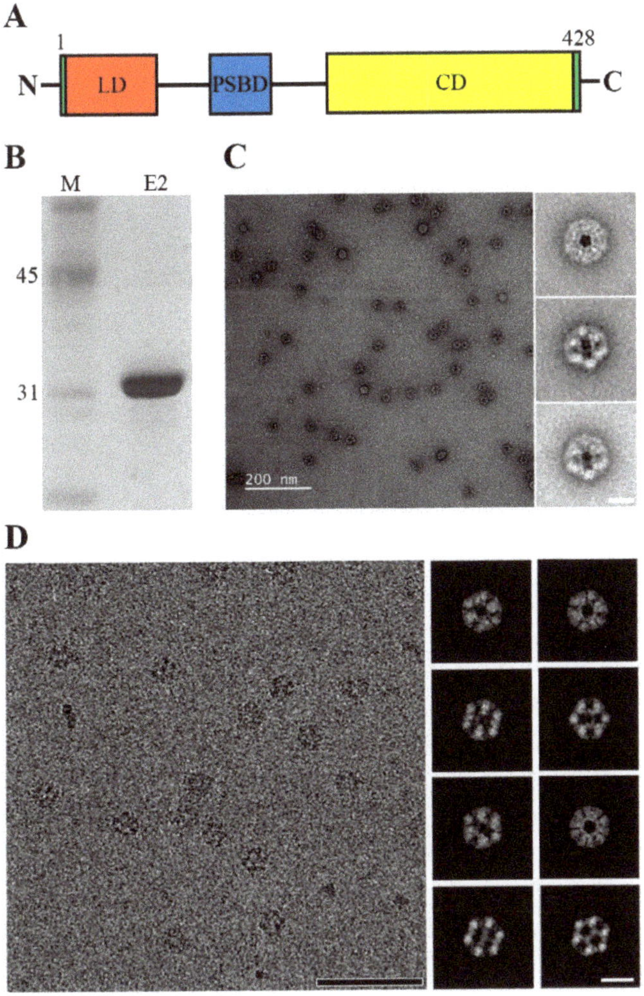

Figure 1. Electron micrographs of the purified E2 complex. (**A**) Composition of the monomeric domain of the E2 inner core complex of the PDC. N, N-terminus; C, C-terminus; LD, lipoyl binding domain; PSBD, peripheral subunit binding domain; CD, core domain. (**B**) Proportion of the purified supernatant E2 complex in SDS-PAGE. M, Marker (31 and 45 kDa). (**C**) Fields of negatively stained (**left**) and selected 2D averages (**right**) of the purified E2. (**D**) Cryo-EM micrograph (**left**) and selected final 2D averages (**right**) from classified particles of the frozen hydrated E2 complex. Scale bars in (**C**), 200 nm (**left**) and 25 nm (**right**); (**D**), 200 nm (**left**) and 20 nm (**right**).

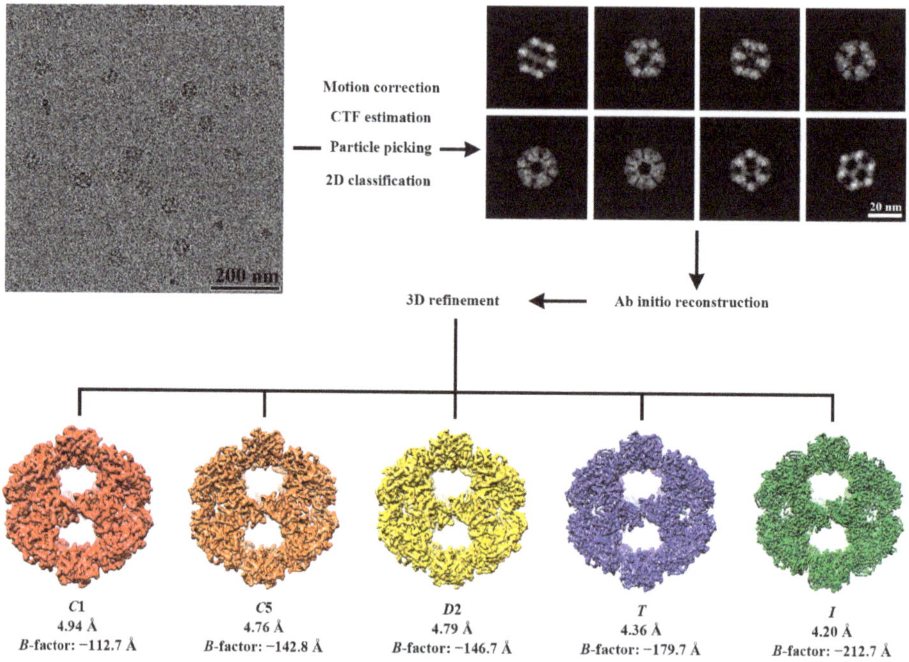

Figure 2. Workflow of cryo-EM data processing. Summarized workflow of data pre-processing (motion correction, CTF estimation, and particle picking), 2D classification, and 3D refinement for each reconstruction.

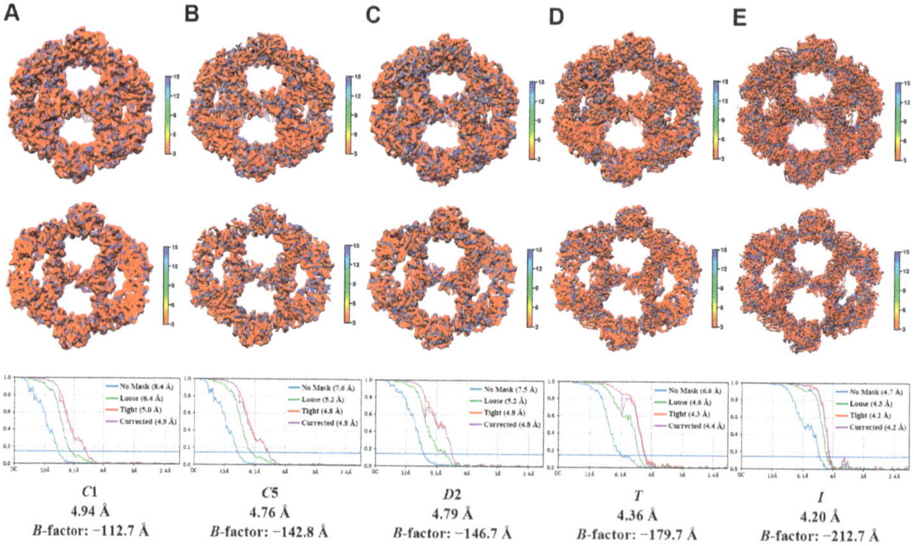

Figure 3. Local and global resolution estimates for cryo-EM maps. (**A**), C1; (**B**), C5; (**C**), D2; (**D**), T; (**E**), I, respectively. Local resolution estimations are illustrated by variation in surface color (top and middle) and were calculated using CryoRes. Global resolution (bottom) was estimated at FSC = 0.143 from half maps and were estimated using CryoSPARC.

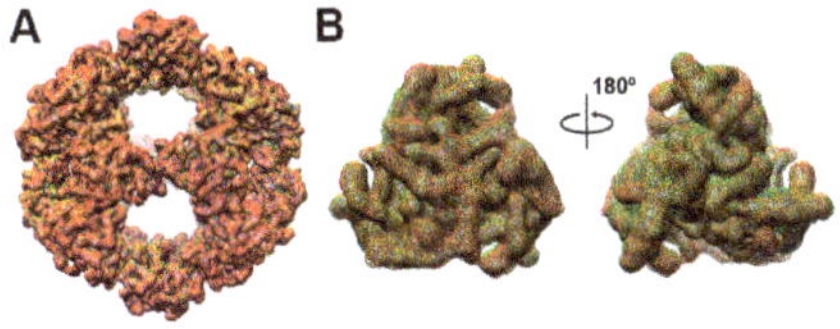

Figure 4. Illustrated comparison of reconstructed cryo-EM maps derived from different symmetric restraints. (**A**) Overall deviations among the five superposed 3D densities, represented by the mesh envelope, corresponding to each symmetric restraint: C1 (red), C5 (orange), D2 (yellow), T (blue), and I (green). (**B**) Enlarged view of the segmented region of each trimeric unit from each model that constitutes the respective map shown (**A**).

2.3. Comparison of the Pseudo-Atomic Models of the E2 Complex with C1 Cryo-EM Map

Next, we conducted a more detailed comparison by performing atomic model building using each model, except for the C1 map. Atomic models of C5, D2, T, and I were built using ModelAngelo [26], which employs sequence-based machine learning to reconstruct each model (Figure 5A–D). The structure of each atomic model corresponds well with the cryo-EM map of C1, which serves as crucial evidence that the individual reconstructions, including that of C1, are similar structures (Figure 6A). Furthermore, to determine whether structural differences exist among the different models (excluding C1), we superposed the different models based on the C1 cryo-EM map (Figure 6B). This result confirmed that the respective atomic models did not structurally differ from one another. Subsequently, to compare with previously published results, we examined the structural differences by fitting our pseudo-atomic model with the highest resolution, I, with the model from PDB-1B5S, onto the C1 map (Figure 6C). This demonstrates that there are no significant structural differences.

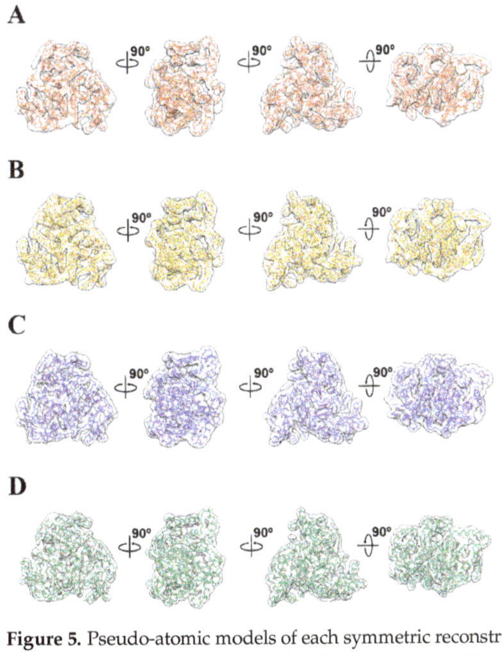

Figure 5. Pseudo-atomic models of each symmetric reconstruction. (**A–D**) Atomic models calculated from each trimeric EM map for detailed comparison. The models for C5, D2, T, and I are colored orange, yellow, blue, and green, respectively. Each atomic model is shown by fitting the envelope of the corresponding EM envelope. The model is rotated by 90° around the x- and y-axes in (**A–D**), respectively.

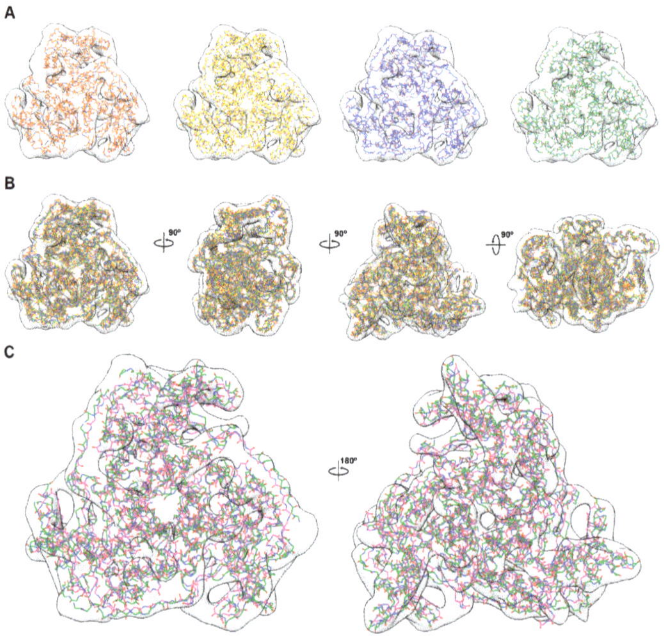

Figure 6. Comparison of each atomic model with the C1 cryo-EM map. (**A**) Shows the superposition of each pseudo-atomic model excluding C1 on to the C1 trimeric complex form of cryo-EM map. (**B**) Combined pseudo-fitted results of all calculated models onto the C1 cryo-EM envelope for comprehensive comparison. The model is rotated by 90° around the x- and y-axes. (**C**) Comparison of the pseudo-fitting of the atomic model generated by I symmetric restrains and the trimeric EM envelope of C1 with the PDB: 1B5S. The respective models are colored as follows: C5, orange; D2, yellow; T, blue; I, green; 1B5S, magenta.

3. Discussion

Deciphering the molecular symmetry in cryo-EM is indeed valuable, especially in SPA, a widely adopted approach for the structural analysis of biological samples. Thus, reliable structural analysis through SPA can provide a deeper understanding of the physiological phenomena such as PPIs and the physiological meaning of these interactions. It offers informative insight into protein assembly resulting from the structural dynamics of proteins. Moreover, the dynamics involved in protein assembly and interaction are critical for their functionality and can influence how proteins respond to environmental changes.

Thus, we compared the results of the E2 complex by applying parameters, such as I, C5, D2, T, and C1. However, due to the 3D models corresponding to a resolution of 4.20 to 4.93 Å, detailed structural analysis of the E2 complex at the atomic level was not possible. Next, although our results consistently reveal structurally identical 3D maps, it has been observed that the resolution gradually decreases compared to I. If the E2 complex adheres perfectly to icosahedral symmetry, employing C1, C5, D2, and T as symmetry parameters in the context of cryo-EM is essentially equivalent to a reduction in the number of particle images of the E2 complex because C1, C5, D2, and T are in fact sub-groups of I symmetry [27,28]. Thus, when considering I encompasses 60 elements, C1, comprising only 1 element, results in a 1/60 reduction. D2, consisting of 4 elements, leads to a 1/15 reduction. C5, with 5 elements, corresponds to a 1/12 reduction. Nevertheless, we confirmed through pseudo-atomic modeling that, despite the use of different parameters, all exhibited the same 3D structure in terms of the structural aspect.

Therefore, a lack of understanding or inexperience with the parameters used for processing could result in vital aspects being overlooked and, ultimately, flawed results. This approach helps to mitigate the risk of overlooking crucial aspects from a biological perspective. Information shown in this study can convey an important message to researchers who study high-resolution protein structures and the structural dynamics of a protein complex.

4. Materials and Methods

4.1. Protein Expression and Purification

The deoxyribonucleic acid (DNA) encoding E2 protein Bacillus stearothermophilus (*B. stearothermophilus*) was amplified by polymerase chain reaction (PCR) using suitable primers and the corresponding cDNA (UniProt: P11961) as templates, and the PCR products were digested and inserted into the *pET 30a* (+) vector (Merck Millipore, Darmstadt, Germany), which had been digested with the corresponding restriction enzymes. The forward and reverse primers used were P1 and P2 for SpyTag-PDH, P1 (BamHI site underlined): 5′-CGC GGA TCC GCT CAT ATT GTT ATG-3′, P2 (XhoI site underlined): 5′-CGG CTC GAG TTA TTA ATA ATA ATT-3′. The prepared DNA was transformed into *Escherichia coli* (*E. coli*) One ShotTM BL 21 StarTM (DE3) (Thermo fisher Scientific, Waltham, MA, USA). The expression of recombinant E2 was induced by adding 0.1 mmol/L isopropyl-β-D-thiogalactoside (IPTG) at 18 °C for 16 h [29]. Cells were harvested by centrifugation (2808× *g* at 4 °C for 15 min), washed with a buffer (50 mM Tris pH 7.5, 500 mM NaCl, 5% glycerol, and 1 mM phenylmethylsulfonyl fluoride (PMSF)), and lysed by ultra-sonication. After centrifugation (30,438× *g* at 4 °C for 25 min), the supernatant was purified by heat precipitation for 20 min at 70 °C. After heating, samples were centrifuged (30,438× *g* at 4 °C for 25 min) [22,30].

4.2. Conventional Transmission Electron Microscopy

The condition of the purified proteins was confirmed with heavy-metal staining. A 5 μL sample of the E2 complex (100 nM) was applied to freshly glow-discharged (Harric Plasma, Ithaca, NY, USA) carbon-coated copper grids. The grids were then negatively stained with 1% uranyl acetate [31] and examined under a JEM-2100F electron microscope (JEOL, Tokyo, Japan) operated at 200 kV. Images were captured with a 4K Gatan Oneview CMOS camera at a nominal magnification of 40,000 (0.28 nm/pixel). Then, the 2D class averages were generated from 1887 particles using the EMAN2 package to enhance the signal-to-noise ratio (SNR) of the negatively stained data for visual inspection [32].

4.3. Cryo-EM Sample Preparation

A sample (4 μL of 100 nM) was applied to a holey carbon grid (Quantifoil R2/R2 100 mesh, Au NH2 finder, EMS, Santa Fe Springs, CA, USA), which had been negatively glow-discharged at 15 mA for 1 min using easiGlow (PELCO, Fresno, CA, USA). The grid was blotted for 5 s with force −1 Ted Pella 595 filter paper (Ted pella, Redding, CA, USA) using Vitrobot Mark IV (Thermo Fisher Scientific, USA) at 4 °C with 100% humidity and then quickly plunged frozen in liquid ethane cooled by liquid nitrogen. Instrumentation was used at the Kangwon Center for Systems Imaging (Chuncheon, Republic of Korea) and Korea Basic Science Institute (Daejeon, Republic of Korea).

4.4. Data Collection and Image Processing

Automated data collection was performed using EPU on a Talos Artica G2 transmission electron microscope (Thermo Fisher Scientific, USA) operating at 200 kV, equipped with a K3 detector. Each movie was recorded with a total dose of 40 e-/Å-2 per movie and a defocus range of −1.8 to −2.6 μm. The movies were collected in electron counting mode with a pixel size of 1.1 Å. A total of 5250 movies were acquired, and before conducting the 3D reconstruction, motion correction was estimated, and the contrast transfer function (CTF) was adjusted. After estimation of the 5250 micrographs, the particles were picked from micrographs without templates (Blob picking). A total of 90,756 particles were extracted. Of these, 69,946 particles were then selected based on visual inspection from the best-matched classes resulting from 2D classification. Next, five initial reference models were created with heterogeneous ab initio reconstruction. Subsequently, using the same initial volume, homogeneous 3D refinement was conducted with the C1, C5, D2, T, and I symmetry parameters. The final model of each symmetry parameter was refined with a soft-edge mask and sharpened with B-factors of −112.7, −142.8, −146.7, −179.7, and −212.7 Å, respectively. All image processing procedures were carried out using CryoSPARC [33]. The resolution of the final cryo-EM map was estimated by gold-standard Fourier shell correlation (FSC) (0.143 criterion) between the two halves of the dataset within CryoSPARC. The local resolution was calculated with CryoRes, a deep learning-based direct estimation [34]. The processing procedure is described in Figure 2, and the data collection and refinement statistics for the cryo-EM are summarized in Table 1. The final maps were deposited with the Electron Microscopy Data Bank (EMDB) (EMD-38333, C1; EMD-38334, C5; EMD-38335, D2; EMD-38340, T; EMD-38336, I;).

Table 1. Data collection and refinement statistics.

Data Collection					
Dataset	**C1 EMD-38333**	**C5 EMD-38334**	**D2 EMD-38335**	**TEMD-38340**	**IEMD-38336**
Particles	69,946	69,946	69,946	69,946	69,946
Pixel size (Å)	1.1	1.1	1.1	1.1	1.1
Defocus range (μm)	−1.5 to −2.6	−1.5 to −2.6	−1.5 to −2.6	−1.5 to −2.6	−1.5 to −2.6
Total electron dose (e-/Å^2)	40	40	40	40	40
Refinement					
Resolution (Å)	4.93	4.76	4.79	4.36	4.20
Map-sharpening B-factor (Å)	−112.7	−142.8	−146.7	−179.7	−212.7
Average B-factor (Å)	−132.7	−162.8	−166.7	−199.7	−232.7
RMSD					
Bond length (Å)		0.155	0.156	0.118	0.109
Bond angle (°)		18.492	19.857	12.514	11.822
Ramachandran plot					
Favored (%)		43.31	41.57	67.23	70.75
Allowed (%)		19.48	20.90	16.29	13.19
Disallowed (%)		37.21	37.53	16.48	16.06
MolProbity score		5.12	5.23	4.65	4.55

4.5. Model Building

A pseudo-atomic model of each model was built using the automated model-building tool, modelAngelo [26]. Based on the map densities and sequence, the sharpened EM maps were segmented into trimeric complexes using UCSF Chimera [35], and each atomic model was automatically calculated using machine learning based on the sequence of PDB-1B5S. The modeling statistics were validated using PHENIX [36] (Table 1), and all structural analyses were conducted using UCSF Chimera.

Author Contributions: Conceptualization, H.-u.K. and H.S.J.; methodology, M.S.J., M.Y.A., Y.H.P., S.H.P., S.J.C., Y.-s.Y., S.J. and Y.K.K.; software, H.-u.K. and H.S.J.; validation, H.-u.K., M.S.J. and H.S.J.; formal analysis, H.-u.K. and M.S.J.; investigation, H.-u.K., M.S.J. and M.Y.A.; resources, S.J.C. and H.S.J.; data curation, H.-u.K.; writing—original draft preparation, H.-u.K. and M.S.J.; writing—review and editing, H.-u.K. and H.S.J.; visualization, H.-u.K. and M.S.J.; supervision, H.S.J. and Y.K.K.; project administration, H.S.J.; funding acquisition, H.S.J. All authors have read and agreed to the published version of the manuscript.

Funding: This research was supported by a grant from the Korea Basic Science Institute (KBSI) National Research Facilities & Equipment Center (NFEC) funded by the Korea government (Ministry of Education) (2019R1A6C1010006 to HSJ), and the National Research Foundation of Korea (NRF) grant funded by the Korea government (MSIT) (2021R1A2C1009404 to HSJ).

Institutional Review Board Statement: Not applicable.

Informed Consent Statement: Not applicable.

Data Availability Statement: The datasets used and/or analyzed during the current study are available from the corresponding author upon reasonable request.

Conflicts of Interest: Authors S.H.P. and S.J.C. were employed by the company AbTis Co. Ltd. The remaining authors declare that the research was conducted in the absence of any commercial or financial relationships that could be construed as a potential conflict of interest.

References

1. Amunts, A.; Brown, A.; Bai, X.C.; Llacer, J.L.; Hussain, T.; Emsley, P.; Long, F.; Murshudov, G.; Scheres, S.H.W.; Ramakrishnan, V. Structure of the yeast mitochondrial large ribosomal subunit. *Science* **2014**, *343*, 1485–1489. [CrossRef] [PubMed]
2. Kuhlbrandt, W. Biochemistry. The resolution revolution. *Science* **2014**, *343*, 1443–1444. [CrossRef] [PubMed]
3. Kim, H.U.; Jung, H.S. Cryo-EM as a powerful tool for drug discovery: Recent structural based studies of SARS-CoV-2. *Appl. Microsc.* **2021**, *51*, 13. [CrossRef] [PubMed]
4. Kim, H.-U.; An, M.Y.; Jung, H.S. Cryo-electron tomography: A triumphant breakthrough in structural biology. *Biodesign* **2023**, *11*, 33–38. [CrossRef]
5. Chua, E.Y.D.; Mendez, J.H.; Rapp, M.; Ilca, S.L.; Tan, Y.Z.; Maruthi, K.; Kuang, H.; Zimanyi, C.M.; Cheng, A.; Eng, E.T.; et al. Better, Faster, Cheaper: Recent Advances in Cryo-Electron Microscopy. *Annu. Rev. Biochem.* **2022**, *91*, 1–32. [CrossRef]
6. Renaud, J.P.; Chari, A.; Ciferri, C.; Liu, W.T.; Remigy, H.W.; Stark, H.; Wiesmann, C. Cryo-EM in drug discovery: Achievements, limitations and prospects. *Nat. Rev. Drug Discov.* **2018**, *17*, 471–492. [CrossRef] [PubMed]
7. Lyumkis, D. Challenges and opportunities in cryo-EM single-particle analysis. *J. Biol. Chem.* **2019**, *294*, 5181–5197. [CrossRef] [PubMed]
8. Shoemaker, S.C.; Ando, N. X-rays in the Cryo-Electron Microscopy Era: Structural Biology's Dynamic Future. *Biochemistry* **2018**, *57*, 277–285. [CrossRef]
9. Kim, H.-U.; Jeong, H.; Chung, J.M.; Jeoung, D.; Hyun, J.; Jung, H.S. Comparative analysis of human and bovine thyroglobulin structures. *J. Anal. Sci. Technol.* **2022**, *13*, 25. [CrossRef]
10. Robertson, M.J.; Meyerowitz, J.G.; Skiniotis, G. Drug discovery in the era of cryo-electron microscopy. *Trends Biochem. Sci.* **2022**, *47*, 124–135. [CrossRef] [PubMed]
11. Costa, T.R.D.; Ignatiou, A.; Orlova, E.V. Structural Analysis of Protein Complexes by Cryo Electron Microscopy. *Methods Mol. Biol.* **2017**, *1615*, 377–413. [PubMed]
12. Plaxco, K.W.; Gross, M. Protein complexes: The evolution of symmetry. *Curr. Biol.* **2009**, *19*, R25–R26. [CrossRef] [PubMed]
13. Goodsell, D.S.; Olson, A.J. Structural symmetry and protein function. *Annu. Rev. Biophys. Biomol. Struct.* **2000**, *29*, 105–153. [CrossRef]
14. Elmlund, D.; Elmlund, H. Cryogenic electron microscopy and single-particle analysis. *Annu. Rev. Biochem.* **2015**, *84*, 499–517. [CrossRef]
15. Zhang, X.; Settembre, E.; Xu, C.; Dormitzer, P.R.; Bellamy, R.; Harrison, S.C.; Grigorieff, N. Near-atomic resolution using electron cryomicroscopy and single-particle reconstruction. *Proc. Natl. Acad. Sci. USA* **2008**, *105*, 1867–1872. [CrossRef] [PubMed]
16. Nakane, T.; Kotecha, A.; Sente, A.; McMullan, G.; Masiulis, S.; Brown, P.; Grigoras, I.T.; Malinauskaite, L.; Malinauskas, T.; Miehling, J.; et al. Single-particle cryo-EM at atomic resolution. *Nature* **2020**, *587*, 152–156. [CrossRef] [PubMed]
17. Vilas, J.L.; Carazo, J.M.; Sorzano, C.O.S. Emerging Themes in CryoEM horizontal line Single Particle Analysis Image Processing. *Chem. Rev.* **2022**, *122*, 13915–13951. [CrossRef] [PubMed]
18. Vilas, J.L.; Tabassum, N.; Mota, J.; Maluenda, D.; Jimenez-Moreno, A.; Majtner, T.; Carazo, J.M.; Acton, S.T.; Sorzano, C.O.S. Advances in image processing for single-particle analysis by electron cryomicroscopy and challenges ahead. *Curr. Opin. Struct. Biol.* **2018**, *52*, 127–145. [CrossRef]
19. Scapin, G.; Potter, C.S.; Carragher, B. Cryo-EM for Small Molecules Discovery, Design, Understanding, and Application. *Cell Chem. Biol.* **2018**, *25*, 1318–1325. [CrossRef] [PubMed]
20. Guaita, M.; Watters, S.C.; Loerch, S. Recent advances and current trends in cryo-electron microscopy. *Curr. Opin. Struct. Biol.* **2022**, *77*, 102484. [CrossRef] [PubMed]
21. Yu, X.; Hiromasa, Y.; Tsen, H.; Stoops, J.K.; Roche, T.E.; Zhou, Z.H. Structures of the human pyruvate dehydrogenase complex cores: A highly conserved catalytic center with flexible N-terminal domains. *Structure* **2008**, *16*, 104–114. [CrossRef] [PubMed]
22. Jiang, J.; Baiesc, F.L.; Hiromasa, Y.; Yu, X.; Hui, W.H.; Dai, X.; Roche, T.E.; Zhou, Z.H. Atomic Structure of the E2 Inner Core of Human Pyruvate Dehydrogenase Complex. *Biochemistry* **2018**, *57*, 2325–2334. [CrossRef] [PubMed]
23. de Marco, A.; Casatta, E.; Savaresi, S.; Geerlof, A. Recombinant proteins fused to thermostable partners can be purified by heat incubation. *J. Biotechnol.* **2004**, *107*, 125–133. [CrossRef]

24. Liu, S.; Xia, X.; Zhen, J.; Li, Z.; Zhou, Z.H. Structures and comparison of endogenous 2-oxoglutarate and pyruvate dehydrogenase complexes from bovine kidney. *Cell Discov.* **2022**, *8*, 126. [CrossRef] [PubMed]
25. Forsberg, B.O.; Aibara, S.; Howard, R.J.; Mortezaei, N.; Lindahl, E. Arrangement and symmetry of the fungal E3BP-containing core of the pyruvate dehydrogenase complex. *Nat. Commun.* **2020**, *11*, 4667. [CrossRef] [PubMed]
26. Jamali, K.; Käll, L.; Zhang, R.; Brown, A.; Kimanius, D.; Scheres, S.H. Automated model building and protein identification in cryo-EM maps. *Nature* **2024**, *628*, 450–457. [CrossRef] [PubMed]
27. Bahadur, R.P.; Chakrabarti, P.; Janin, J. Protein–protein interaction and quaternary structure. *Q. Rev. Biophys.* **2008**, *41*, 133–180.
28. Yeates, T.O.; Liu, Y.; Laniado, J. The design of symmetric protein nanomaterials comes of age in theory and practice. *Curr. Opin. Struct. Biol.* **2016**, *39*, 134–143. [CrossRef]
29. Kim, H.-u.; Park, Y.H.; An, M.Y.; Kim, Y.K.; Song, C.; Jung, H.S. Structural insights into calcium-induced conformational changes in human gelsolin. *Biochem. Biophys. Res. Commun.* **2024**, *735*, 150826. [CrossRef]
30. Izard, T.; Aevarsson, A.; Allen, M.D.; Westphal, A.H.; Perham, R.N.; de Kok, A.; Hol, W.G. Principles of quasi-equivalence and Euclidean geometry govern the assembly of cubic and dodecahedral cores of pyruvate dehydrogenase complexes. *Proc. Natl. Acad. Sci. USA* **1999**, *96*, 1240–1245. [CrossRef]
31. Kim, H.-u.; Bharda, A.V.; Moon, J.C.; Jeoung, D.; Chung, J.M.; Jung, H.S. Microscopic studies on severing properties of actin-binding protein: Its potential use in therapeutic treatment of actin-rich inclusions. *J. Anal. Sci. Technol.* **2021**, *12*, 49. [CrossRef]
32. Tang, G.; Peng, L.; Baldwin, P.R.; Mann, D.S.; Jiang, W.; Rees, I.; Ludtke, S.J. EMAN2: An extensible image processing suite for electron microscopy. *J. Struct. Biol.* **2007**, *157*, 38–46. [CrossRef] [PubMed]
33. Punjani, A.; Rubinstein, J.L.; Fleet, D.J.; Brubaker, M.A. cryoSPARC: Algorithms for rapid unsupervised cryo-EM structure determination. *Nat. Methods* **2017**, *14*, 290–296. [CrossRef] [PubMed]
34. Dai, M.; Dong, Z.; Xu, K.; Zhang, Q.C. CryoRes: Local Resolution Estimation of Cryo-EM Density Maps by Deep Learning. *J. Mol. Biol.* **2023**, *435*, 168059. [CrossRef] [PubMed]
35. Pettersen, E.F.; Goddard, T.D.; Huang, C.C.; Couch, G.S.; Greenblatt, D.M.; Meng, E.C.; Ferrin, T.E. UCSF Chimera—A visualization system for exploratory research and analysis. *J. Comput. Chem.* **2004**, *25*, 1605–1612. [CrossRef]
36. Liebschner, D.; Afonine, P.V.; Baker, M.L.; Bunkoczi, G.; Chen, V.B.; Croll, T.I.; Hintze, B.; Hung, L.W.; Jain, S.; McCoy, A.J.; et al. Macromolecular structure determination using X-rays, neutrons and electrons: Recent developments in Phenix. *Acta Crystallogr. D Struct. Biol.* **2019**, *75 Pt 10*, 861–877. [CrossRef]

International Journal of
Molecular Sciences

MDPI

Article

Structural Implications of H233L and H398P Mutations in Phospholipase Cζ: A Full-Atom Molecular Dynamics Study on Infertility-Associated Dysfunctions

Fernando Hinostroza [1,2,3,4,*], Sofía Albornoz-Muñoz [5], Sebastián Vergara [3], Gabriela Urra [6], Ingrid Araya-Durán [7], Rafael A. Fissore [8], Fernando Danilo González-Nilo [7], Daniel Bustos [6] and Ingrid Carvacho [3,*]

1 Centro de Investigación de Estudios Avanzados del Maule (CIEAM), Vicerrectoría de Investigación y Postgrado, Universidad Católica del Maule, Talca 3460000, Chile
2 Centro de Investigación en Neuropsicología y Neurociencias Cognitivas (CINPSI Neurocog), Facultad de Ciencias de la Salud, Universidad Católica del Maule, Talca 3460000, Chile
3 Departamento de Medicina Traslacional, Facultad de Medicina, Universidad Católica del Maule, Talca 3460000, Chile; sebadiantian@gmail.com
4 Centro para la Investigación Traslacional en Neurofarmacología, Universidad de Valparaíso, Valparaíso 2340000, Chile
5 Escuela de Ingeniería en Biotecnología, Facultad de Ciencias Agrarias y Forestales, Universidad Católica del Maule, Talca 3460000, Chile; sofiaalbornozhb@gmail.com
6 Laboratorio de Bioinformática y Química Computacional, Departamento de Medicina Traslacional, Facultad de Medicina, Universidad Católica del Maule, Talca 3460000, Chile; gabriela.urra@alu.ucm.cl (G.U.); dbustos@ucm.cl (D.B.)
7 Center for Bioinformatics and Integrative Biology (CBIB), Universidad Andrés Bello, Santiago 8370146, Chile; ingrid.araya.duran@gmail.com (I.A.-D.); danilo.gonzaleznilo@gmail.com (F.D.G.-N.)
8 Department of Veterinary and Animal Sciences, University of Massachusetts, Amherst, MA 01003, USA; rfissore@vasci.umass.edu
* Correspondence: fhinostroza@ucm.cl (F.H.); icarvacho@ucm.cl (I.C.)

Abstract: Phospholipase Cζ (PLCζ), a sperm-specific enzyme, plays a critical role in mammalian fertilization. Mutations in PLCζ have been linked to male infertility, as they impair its ability to trigger calcium (Ca^{2+}) oscillations necessary for egg activation and embryo development. During fertilization, PLCζ is introduced into the egg, where it hydrolyzes phosphatidylinositol 4,5-bisphosphate (PIP_2) into inositol 1,4,5-trisphosphate and diacylglycerol, leading to Ca^{2+} release from the endoplasmic reticulum. Human infertility-associated mutations include H233L, H398P, and R553P, which disrupt PLCζ function. To elucidate the molecular consequences of the mutations, we employed full-atom molecular dynamics simulations to analyze structural perturbations and their impact on PIP_2 and Ca^{2+} binding. Our results reveal that H233L and H398P mutations significantly reduce interactions with PIP_2, disrupting hydrogen bonding and salt bridge formation, leading to misalignment of the substrate. Additionally, these mutations destabilize Ca^{2+} binding by altering its positioning within the active site. In contrast, the R553P mutation primarily affects intramolecular stability and enzyme dynamics without impairing substrate or ion binding. Free energy calculations indicate an increased affinity for PIP_2 in H233L and H398P mutants, leading to an aberrant substrate positioning and compromised hydrolysis. These structural insights help explain the egg activation failure and infertility of patients carrying these mutations.

Keywords: Phospholipase Cζ; phosphatidylinositol 4,5-bisphosphate; infertility; H233L mutation; H398P mutation; R553P mutation

1. Introduction

Infertility affects millions of men and women of reproductive age. Worldwide, infertility affects ~15% of couples, and 7% of males suffer from infertility. Male infertility accounts for ~30% of the total infertility cases, and genetic defects cause male infertility, including mutations in PLCZ1 [1–3], which encodes Phospholipase Cζ (PLCζ).

PLCζ (PLCZ1) is a sperm-specific enzyme that triggers calcium (Ca^{2+}) oscillations and egg activation [4,5]. Following release into the egg upon fusion, PLCζ hydrolyzes phosphatidylinositol 4,5-bisphosphate (PIP_2), associated with cellular membranes and vesicles, producing inositol 1,4,5-triphosphate (IP_3) [4,6,7]. IP_3 binds IP_3 receptors, the most common in eggs, in type 1 (IP3R1), inducing Ca^{2+} release from the endoplasmic reticulum (ER) and periodic intracellular Ca^{2+} concentration (Ca^{2+}) elevations, known as Ca^{2+} oscillations [8,9] (Figure 1). Sperm derived from mice lacking PLCζ1 (*Plcz1*$^{-/-}$) fail to induce Ca^{2+} oscillations [10] or show an impaired pattern of them with a decreased number of oscillations compared with WT sperm [11]. *Plcz1*$^{-/-}$ sperm fertilization results in polyspermy when fertilizing eggs in vivo. Consequently, these findings confirm the essential role of the Ca^{2+} oscillations in modulating the block of polyspermy [10]. The fertility of *Plcz1*$^{-/-}$ males is also greatly affected, and the animals exhibit a subfertile phenotype. However, *Plcz1*$^{-/-}$ animals are not sterile, suggesting a PLCζ1-independent fertility route [10,11]. Moreover, PLCζ mutations identified in a cohort of couples undergoing fertility treatment were replicated in mouse models to assess their impact on early embryonic cleavage and pregnancy outcomes. The authors reported lower rates of cell division, impaired embryo quality, lower pregnancy success, and smaller litter sizes following IVF using sperm from *Plcζ* mutant males with WT eggs, compared to IVF with WT *Plcζ* sperm [12]. Therefore, the PLCζ function is required for the egg-to-embryo transition. In fact, suppression or failure to initiate Ca^{2+} oscillations causes egg activation failure and impairs early embryo development [13,14].

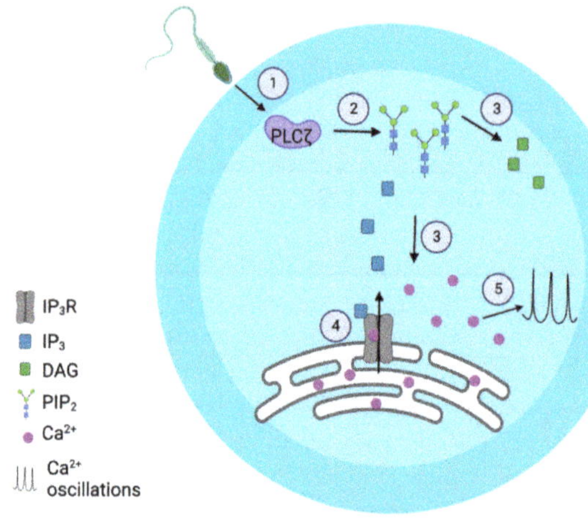

Figure 1. PLCζ induces Ca^{2+} oscillations in the egg. 1. Sperm fuses with the egg and releases its genetic material and the PLCζ into the egg. 2. PLCζ binds PIP_2. 3. PLCζ hydrolyzes PIP_2 to produce IP_3 and DAG. 4. IP_3 binds IP_3 receptors in the endoplasmic reticulum and induces Ca^{2+} release. 5. The release of intracellular Ca^{2+} triggers Ca^{2+} oscillations and activates the egg.

Structurally, PLCζ has two EF-hand domain pairs in its N-termini, a catalytic domain composed of X and Y domains connected by a loop called X-Y linker, and a C2 domain in its C-termini region [4,15] (Figure 2A,B). Unlike other cellular PLCs, PLCζ lacks an

N-terminus PH domain. The integrity of the four EF-hand domains and the C-terminal C2 domain is essential for PLCζ to reach its maximal activity [16], while the catalytic domain is involved in PIP_2 hydrolysis facilitated by Ca^{2+} binding [15,17]. The X-Y linker is proposed to interact with the plasma membrane (PM) and PIP_2 [17,18].

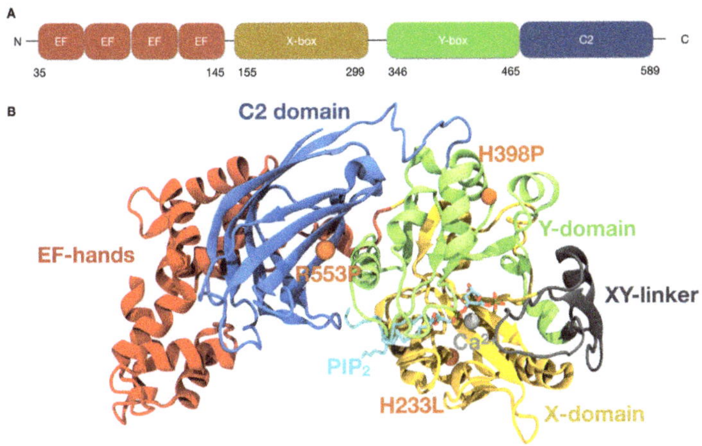

Figure 2. PLCζ structure. (**A**) The N-terminus of the PLCζ has an EF-hand domain, and X- and Y-domains linked by an XY linker, and a C2 domain in its C-terminus. (**B**) AlphaFold model of the human PLCζ binding PIP_2 and Ca^{2+}.

Several mutations in PLCζ have been related to male infertility [2,3,19,20]. They are generally located in the catalytic domain (X or Y), but some have been identified in the C2 domain. In fact, the mutation I489F located in the C2 domain is linked to infertility in humans. In vitro studies demonstrated that microinjection of the mutant protein at physiological concentrations failed to induce Ca^{2+} oscillations or support embryo development. Interestingly, the phenotype could be rescued by increasing the concentration of the recombinant I489F mutant protein. Biochemical analyses revealed that the mutation does not impair enzymatic activity or the Ca^{2+} sensitivity of the PLCζ. However, I489F significantly disrupts the binding of PLCζ to PI(3)P and PI(5)P, underlying the infertility phenotype reported in the patient [21]. Only one of the mutations is in the EF domain [5]. The mutations H233L and H398P, which comprised exchanges of histidine for leucine in position 233 [2] (Figure 2B) and for proline in position 398 [13]. Sperm from patients carrying the mutation H233L were injected into human eggs, and oocyte activation was evaluated, showing a lower percentage of activation than WT oocytes [20]. PLCζ1 mutation H398L patients are infertile, and in the mouse oocyte activation test (MOAT), injection of the mutated cRNA showed an abnormal pattern of Ca^{2+} oscillations [13]. The mutations were initially found in the same patient as compound heterozygous mutations with maternal and paternal inheritance, respectively. The H233L mutation is located in the X-domain, part of the catalytic site (Figure 2) [2,20], and H398P in the Y-box domain (Figure 2) [13]. This patient also seemed to have reduced expression of PLCζ [1]. A more recent homozygous mutation was described in the PLCζ C2 domain, R553P, where arginine was exchanged for proline (Figure 2). Injection of R553P Plcζ cRNA into mouse oocytes failed to activate eggs and trigger embryo development [22]. Although PLCZ1 mutations are individually rare, their clinical relevance is substantial. Notably, 33.6% of men who experience fertilization failure after intracytoplasmic sperm injection (ICSI) have been found to carry *PLCZ1* variants [23], highlighting the importance of genetic screening in assisted reproduction settings.

Despite the obvious impact on fertility, it remains elusive how these mutations affect the structure of PLCζ and its association with PIP_2 at a molecular level. Here, we performed long-timescale full-atom molecular dynamics (MD) simulations to understand at an atomic level how these mutations located in different domains impact the structure of PLCζ and the binding of its natural ligand, PIP_2. Our data showed that the H233L and H398P displayed fewer hydrogen bonds and salt bridges between PLCζ and PIP_2, reduced total contacts and binding energy, suggesting a destabilization of the PLCζ-PIP_2 complex. Moreover, the simulations indicated that these mutations modified PLCζ Ca^{2+} ion binding. On the other hand, the R553P mutation affects intramolecular dynamics and possibly membrane-binding. In conclusion, our results showed that H233L and H398P mutations modified the structure of PLCζ, triggering a shift in the position of PIP_2 and Ca^{2+}. Therefore, preventing PIP_2 hydrolysis. Our data provide relevant insights into the structure of PLCζ and the mechanisms by which its function is precisely regulated, highlighting its fundamental role in human fertility.

2. Results

2.1. H233L and H398P Mutations Altered PIP_2 Binding

The injection of H233L, H398P, and R553P PLCζ mRNA mutants or the presence of these mutations in human patients fails to activate eggs and causes infertility [2,20,22]. To assess whether this failure alters PIP_2 binding, we measured the Mean Square Deviation (MSD) of PLCζ for PIP_2, which measures the average squared distance that the molecules travel over time within the binding site. The average PIP_2 MSD values for the H233L and H398P mutants were higher than those for R553P and WT (Figure 3A), indicating greater molecular spatial dispersion and instability during the simulated interaction (H233L: 5.11 ± 0.28 Å^2, H398P: 9.11 ± 0.54 Å^2, R553P: 1.80 ± 0.06 Å^2, WT: 2.18 ± 0.09 Å^2). Notably, despite the inherent flexibility of the PIP_2 acyl chains, the overall MSD values in WT and R553P remained low, indicating stable binding within the active site. This suggests that the lateral chain mobility does not significantly affect the MSD measurement, supporting the validity of our approach without requiring a separate analysis of the myo-inositol bisphosphate moiety.

To further comprehend the deviation observed for the mutants H233L and H398P, we calculated the distance of amino acids involved in PIP_2 stability. We observed that the N171, which interacts with the 1-OP4, exhibits an average distance of 4.23 ± 0.003 Å for the WT and 4.38 ± 0.003 Å for the R553P mutant. In contrast, the H233L and the H398P mutants showed a larger distance, being 5.47 ± 0.006 Å and 5.39 ± 0.008 Å, respectively (Figure 3B). The K297, S378, and R405 stabilize the 4-OP4. The average distance for the K297 in the WT enzyme was 6.06 ± 0.08 Å, and 5.77 ± 0.004 Å, 9.11 ± 0.01 Å, and 5.99 ± 0.009 Å for the H233L, H398P, and R553P mutants, respectively (Figure 3C). For the S378, the average distance was 5.88 ± 0.007 Å, 6.3 ± 0.004 Å, 7.8 ± 0.01 Å, and 5.21 ± 0.008 Å for the WT, H233L, H398P, and R553P proteins, respectively (Figure 3D). The average distance for the R405 in the WT and the R553P enzymes was 4.63 ± 0.006 Å and 4.61 ± 0.008 Å. In contrast, the H233L and the H398P mutants exhibited 5.67 ± 0.005 Å and 7.08 ± 0.01 Å (Figure 3E). The K299 contacts the 5-OP4 and in the WT enzyme has an average distance of 9.03 ± 0.01 Å. In contrast, the H233L, H398P, and the R553P mutants 7.98 ± 0.01 Å, 10.16 ± 0.02 Å, and 9.06 ± 0.01 Å, respectively (Figure 3F). These results indicate that the mutation H233L and the H398P in PLCζ impair PIP_2 binding, favoring mispositioning of PIP_2, which would prevent its effective hydrolysis and cause failure to initiate oscillations.

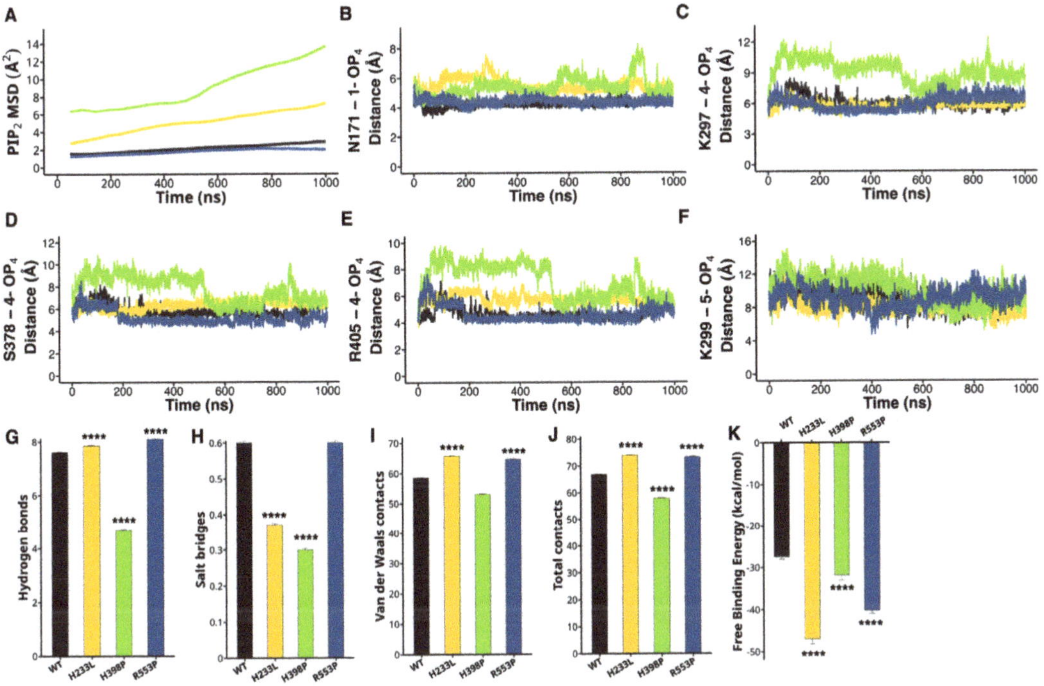

Figure 3. H233L and H398P mutations impair PIP_2 binding. MD of the PLCζ mutants H233L (yellow), H398P (green), and R553P (blue) in comparison with WT (black) are evaluated using structural and functional parameters. (**A**) Mean Square Deviation (MSD) of PIP_2 during the simulation. (**B**) Average distance of N171 with 1-OP4. Distance of (**C**) K297, (**D**) S378, and (**E**) R405 with 4-OP4. (**F**) Average distance of K299 with 5-OP4. Amount of (**G**) hydrogen bonds, (**H**) salt bridges, (**I**) van der Waals, and (**J**) total contacts of PIP_2 with PLCζ. (**K**) Free binding energy of PIP_2 with PLCζ. **** $p < 0.0001$.

We also analyzed the types of contacts between PLCζ and PIP_2, including hydrogen bonds, salt bridges, and van der Waals contacts. In the WT complex, PIP_2 formed an average of 7.6 ± 0.01 hydrogen bonds with WT PLCζ, whereas the H233L complex formed 7.8 ± 0.02 bonds (**** $p < 0.0001$), the H398P complex formed 4.6 ± 0.01 bonds (**** $p < 0.0001$), and the R553P complex formed 8.1 ± 0.01 bonds (**** $p < 0.0001$), all statistically different from the WT (Figure 3G). Salt bridges were less frequent in H233L and H398P (0.37 ± 0.005, 0.3 ± 0.004, respectively, **** $p < 0.0001$), but R553P and WT were comparable (0.6 ± 0.006) (Figure 3H). On the contrary, van der Waals' contacts were more frequent for H233L (65.7 ± 0.1, **** $p < 0.0001$) and R553P mutant (64.5 ± 0.1, **** $p < 0.0001$) compared to H398P (53 ± 0.1) and WT PLCζ (58.5 ± 0.08) (Figure 3I). However, because of the variation in type and number of interactions with PIP_2, we calculated the total bonds formed between PLCζ and PIP_2 for the different PLCζ under consideration. We found that the WT protein had a total of 66.7 ± 0.09 contacts, whereas H233L, H398, and R553P mutants displayed 73.9 ± 0.1, 57.9 ± 0.1, and 73.29 ± 0.1 contacts, respectively (**** $p < 0.0001$; Figure 3J).

Next, we compared the free binding energy of PIP_2 to PLCζ. The WT complex exhibited a mean binding energy of -27.43 ± 0.5 kcal/mol. In contrast, the average binding energy for the H233L, H398P, and R553P was -47 ± 1.1 kcal/mol, -31.8 ± 1.2 kcal/mol, and -40.22 ± 0.7 kcal/mol, respectively (**** $p < 0.0001$; Figure 3K). Lastly, we evaluated the PIP_2 diffusion coefficient and found higher values for the H233L and the H398P mutants vs. the WT and the R553P mutant (Figure S1A).

2.2. H233L and H398P Mutations Impair Ca^{2+} Binding

Ca^{2+} ions play a crucial role in facilitating PIP_2 binding by PLCζ [15]. Therefore, we analyzed whether H233L, H398P, and R553P mutations modify Ca^{2+} binding to PLCζ. Our analysis revealed that the WT PLCζ formed, on average, 3.8 ± 0.005 contacts with Ca^{2+}, whereas the H233L and H398P mutations had fewer amino acid contacts, 3.0 ± 0.008 and 3.4 ± 0.005, respectively. The R553P mutant had an average of 4.1 ± 0.002 contacts with Ca^{2+} throughout the simulation (Figure 4A).

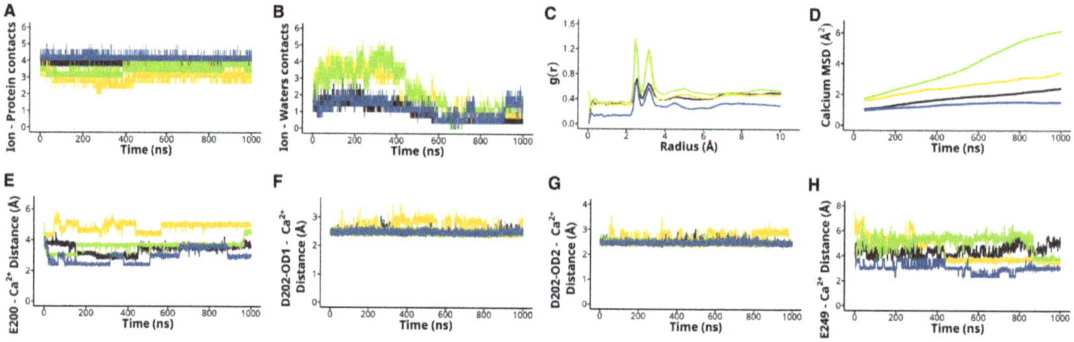

Figure 4. H233L and H398P mutations modify Ca^{2+} binding. The Ca^{2+} binding dynamics for the PLCζ mutants H233L (yellow), H398P (green), and R553P (blue) are analyzed by the evaluation of Ca^{2+} contacts and hydration. (**A**) Amount of Ca^{2+} contacts with PLCζ. (**B**) Amount of Ca^{2+} contacts with water molecules. (**C**) Radial distribution function (g(r)) of water molecules around the Ca^{2+} ion. (**D**) The Ca^{2+} Mean square deviation (MSD) throughout the simulation. Average distance of (**E**) E200, (**F**) D202-OD1, (**G**) D202-OD2, and (**H**) E249 with Ca^{2+}.

We also measured the number of contacts the Ca^{2+} ion made with water molecules. Throughout the simulation, we found that all systems showed a reduction in the number of water contacts, with the WT PLCζ having an average of 1.0 ± 0.006 and the R553P having 1.1 ± 0.007 contacts, whereas the H233L had 2.1 ± 0.01 and H398P had 2.2 ± 0.01 contacts (Figure 4B). Notably, during the first half of the simulation, the H233L and H398P systems established nearly twice the number of solvation contacts with Ca^{2+} compared to the other systems. To complement this analysis, we quantified the density of water molecules around Ca^{2+} by measuring the radial distribution function (RDF, g(r)). The peaks indicate preferred distances where water molecules are more likely to be found, thus reflecting the hydration shell and interaction dynamics between the ion and the surrounding water. We found that the systems had two peaks around 2.55 and 3.15 Å RDF, larger for H233L and H398P than for R553P and the WT protein (Figure 4C).

Further, we found that the MSD of Ca^{2+} throughout the simulation in the WT PLCζ had a displacement of 1.7 ± 0.1 $Å^2$. In contrast, the H233L, H398P, and R553P mutations exhibited MSDs of 2.5 ± 0.07, 3.9 ± 0.2, and 1.3 ± 0.03, respectively (Figure 4D). Consistent with these values, the Ca^{2+} ion's diffusion coefficient for the mutations H233L and H398P showed a higher value than the one for the R553P mutation and the WT complex (Figure S1B). Specifically, the mean diffusion coefficients for H233L (0.000319 ± 0.0000940 $Å^2$/ns) and H398P (0.000443 ± 0.0000485 $Å^2$/ns) were nearly double those of the R553P (0.000173 ± 0.0000688 $Å^2$/ns) and WT (0.000222 ± 0.0000541 $Å^2$/ns) complexes, indicating greater ion mobility in these two mutants.

Additionally, we calculated the distance of E200, D202, and E249, which stabilizes the Ca^{2+}. The average distance of the E200 in the WT enzyme was 3.42 ± 0.004 Å, whereas for the H233L, H398P, and the R553P mutants, it was 4.87 ± 0.004 Å, 3.63 ± 0.004 Å, and 2.92 ± 0.005 Å, respectively (Figure 4E). We also measured the average distance of both oxygens (OD1 and OD2) bonded to the gamma carbon of D202, which interacts with Ca^{2+}. For the D202-OD1, the WT average distance was 2.47 ± 0.001 Å, for the H233L enzyme was 2.62 ± 0.003 Å, for the H398P mutant was 2.46 ± 0.0008 Å, and for the R553P protein was 2.47 ± 0.0008 Å (Figure 4F). For the D202-OD2, the average distance of the WT PLCζ was 2.48 ± 0.001 Å, whereas for the H233L, H398P, and the R553P mutants, it was 2.62 ± 0.003 Å, 2.47 ± 0.0008 Å, and 2.48 ± 0.0008 Å, respectively (Figure 4G). For the E249, the distance was 4.43 ± 0.006 Å for the WT, 4.48 ± 0.01 Å for the H233L mutant, 5.12 ± 0.008 Å for the H398P enzyme, and 3.05 ± 0.005 Å for the R553P protein (Figure 4H). These results suggest that the H233L and H398P mutations modify PLCζ Ca^{2+} ion binding, impairing PIP_2 binding and hydrolysis.

2.3. The H233L, H398P, and R553P Mutations Changes the PLCζ Intramolecular Interactions

Hydrogen bonds are crucial non-bonded interactions that stabilize protein structures [24]. Thus, we calculated the number of hydrogen bonds to determine whether the PLCζ mutations modify the number of hydrogen bonds present in each domain (EF-hands, XY, and C2 domains). The WT EF hands domain had an average of 24.1 ± 0.03 hydrogen bonds, whereas the H233L mutation had 26.6 ± 0.03, the H398P had 26.7 ± 0.03, and the R553P mutant had 23.3 ± 0.03 (Figure 5A). The WT XY domain exhibits an average of 61.4 ± 0.06 hydrogen bonds. In contrast, the H233L, H398P, and R553P mutants had 67.0 ± 0.05, 66.5 ± 0.05, and 59.6 ± 0.05 hydrogen bonds, respectively (Figure 5B). The WT C2 domain showed 19.9 ± 0.03 hydrogen bonds, whereas the H233L, H398P, and the R553P mutants exhibited 21.8 ± 0.03, 22.1 ± 0.03, and 18.3 ± 0.03 hydrogen bonds, respectively (Figure 5C).

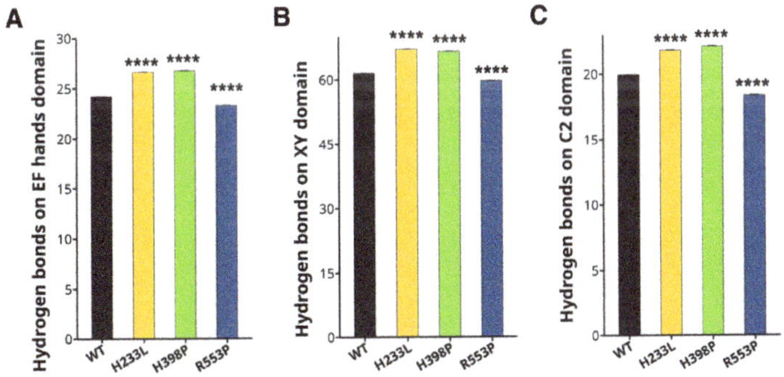

Figure 5. H233L and H398P changes intramolecular hydrogen bonds of PLCζ. Hydrogen bonds are analyzed in the PLCζ mutants H233L (yellow), H398P (green), R553P (blue), and WT (black). Amount of intramolecular hydrogen bonds in the (**A**) EF hands, (**B**) XY, and (**C**) C2 domains. **** $p < 0.0001$.

To explore whether the H233L, H398P, and R553P mutations influence the overall protein structure, we measured the radius of gyration (Rg) of each PLCζ. The Rg measures the compactness and overall shape of the enzyme. We found no significant difference between the WT and the mutant PLCζ in the Rg of the EF-hands domain and C2 domain (Figure 6A,C). On the contrary, the XY domain of the H398P exhibited a bigger Rg during the last 300 ns of simulation (Figure 6B). The Rg of PIP_2 of the WT complex was similar to the H233L and R553P mutants but slightly bigger for the H398P mutant during the last 100 ns of simulation (Figure S2A). Then, we calculated each domain's Root Mean Square Deviations (RMSD) for the four different conditions. RMSD quantifies the enzyme's overall conformational changes and structural stability during the simulation. There was no significant difference between the RMSD of WT compared to the H233L and R553P mutants (Figure 6D,F). In contrast, the XY domain of the H398P mutant showed a bigger RMSD with a mean value of 5.12 ± 1.52 Å, indicating greater structural deviation and instability. This is particularly evident when compared to the WT protein (4.77 ± 0.517 Å), which maintained more consistent structural integrity (Figure 6E). Our data showed that the H233L, H398P, and R553P mutations do not change the overall structure of PLCζ but change the intramolecular interactions mediated by hydrogen bonds.

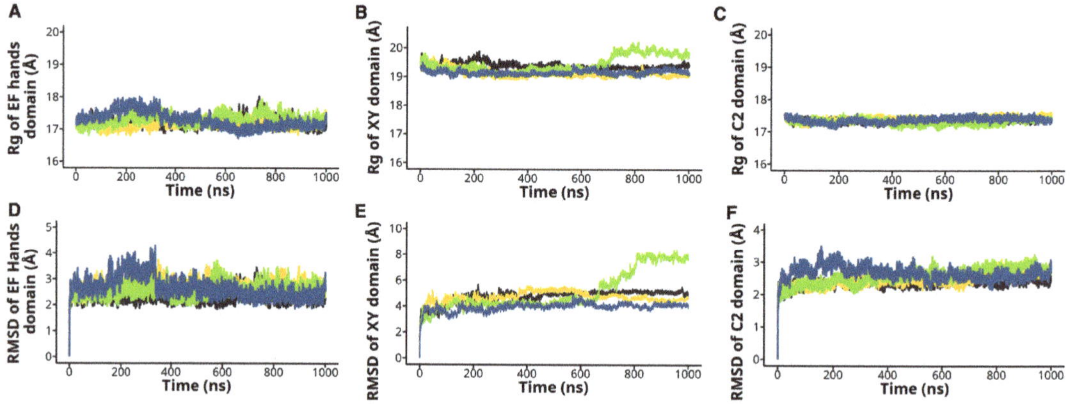

Figure 6. H233L, H398P, and R553P mutations did not affect the overall PLCζ structure. The radius of gyration (Rg) of (**A**) EF hands, (**B**) XY, and (**C**) C2 domains was analyzed for the PLCζ mutants H233L (yellow), H398P (green), and R553P (blue) and WT (black). Root mean square deviation (RMSD) of (**D**) EF hands, (**E**) XY, and (**F**) C2 domains was also determined for the mutants and WT proteins.

3. Discussion

Infertility is a complex condition affecting millions worldwide, with a significant proportion attributed to male factors [25]. Among the genes implicated in male infertility, PLCζ has emerged as a critical player due to its pivotal role in initiating egg activation through Ca^{2+} oscillations during fertilization. However, the molecular mechanism underlying the effects of specific mutations on PLCζ function remains poorly understood. Point mutations in the PLCζ gene can induce structural alterations in the protein, impacting male fertility. Despite their significance, the structural and functional consequences of such mutations have been poorly characterized. In this work, we employed MD simulations to examine the impact of three PLCζ infertility-causing mutations, H233L, H398P, and R553P, on its structure and interactions with its natural ligands PIP_2 and Ca^{2+}. By employing molecular dynamics simulations in an aqueous environment, we aimed to isolate and analyze the intrinsic structural effects of these mutations to further explain infertility in the patients carrying these changes. Our results shed light on the structural alterations induced by these mutations and provide insights into their impact on PLCζ function.

Once the sperm fuses with the egg oolemma, and PLCζ enters the ooplasm, the enzyme hydrolyzes its substrate PIP_2 located in cellular membranes [4,6,7]. The active site structure and main residues are conserved in the PLC family and among mammals, including PLCζ [26]. The structure of PLC∂1's active site, solved by Essen et al., [15] displays 64% identity with the active site of hPLCζ1, indicating a similar PIP_2 binding mode and hydrolysis [4]. PLCζ binds and stabilizes PIP_2 before hydrolysis by interacting with the 4- and 5-phosphates of the inositol group through the amino acids K299, K327, S378, R405, and Y407 and by interacting with the 2-OH and 3-OH groups with the residues D202, E249, and R405. Based on the evidence of increased MSD of PIP_2 and altered protein-ligand contacts, including fewer hydrogen bonds and salt bridges between the enzyme and PIP_2 for the H233L and H398P PLCζ mutants compared to the WT PLCζ, our model predicts impaired PIP_2 binding and a destabilization of the PLCζ-PIP_2 complex.

These structural changes likely interfere with the proper positioning of PIP_2 for hydrolysis, thereby impairing the generation of IP_3 production, Ca^{2+} oscillations, and egg activation.

PLCζ is ~100-fold more sensitive to Ca^{2+} than PLC∂1 and is the most Ca^{2+}-sensitive PLC enzyme [27]. Ca^{2+} plays an important role in the PLCζ-PIP_2 binding since it lowers this interaction's pKa [28,29]. Ca^{+2} is stabilized by the amino acids E200, D202, and E249, and PIP_2 hydrolysis is produced by a nucleophilic attack of H215 on the 1-phosphate, producing DAG and IP_3. In the H233L and H398P mutants, we observed a shift in the positions of PIP_2 and Ca^{2+} in the PLCζ's catalytic site, which is supported by the changes in MSD, the number of amino acids contacting PIP_2 and Ca^{2+}, and the RDF of the water molecules around Ca^{2+}. Surprisingly, the free binding energy of the three mutants is lower than that of the WT, suggesting that the mutant PLCζs bind PIP_2 with more affinity. However, the higher affinity appears to favor incorrect binding, diminishing PIP_2 hydrolysis and, ultimately, preventing Ca^{2+} oscillations. This agrees with the fact that the PLCζ H233L, H398P, and R553P mutations reduce or suppress egg activation in humans [2,20,22]. Remarkably, H233L and H398P are not located within the enzyme's active site but adjacent to it. However, these mutations appear to induce allosteric perturbations that disrupt PIP_2 and Ca^{2+} binding. Specifically, 1-OP4 is displaced from N171, while 4-OP4 is repositioned away from K297, S378, and R405. Additionally, Ca^{2+} is displaced from its stabilizing residues E200 and D202, further compromising enzymatic function. This mutation-induced allosteric changes that are extended through the protein, impacting critical binding sites, have also been observed in other enzymes and proteins [30,31].

Interestingly, the R553P mutation did not significantly affect PIP_2 or Ca^{2+} binding. These results agree that C2 domain deletion does not affect PLCζ enzymatic activity or Ca^{2+} sensitivity [32,33]. This mutation changes the residue cross-correlation patterns, suggesting a disruption of intramolecular interactions within PLCζ. However, the overall protein structure remained largely unaffected by the mutation, as indicated by similar Rg and RMSD values compared to the WT. The C2 domain is proposed to mediate the interaction with phospholipids present in the membrane, facilitating PLCζ's adequate positioning and possible access to substrates [32,33]. Noteworthy, a homozygous I498F mutation in the C2 domain of hPLCζ impaired its distribution in eggs after mRNA injection and in the sperm before fertilization [19]. Thus, it is possible that the observed intramolecular changes could affect the C2 domain amino acids disposition and, subsequently, its binding to membranes. However, the exact mechanism by which PLCζ interacts with membranes remains unclear, highlighting the need for further research to elucidate this process.

Mice deficient in Plcζ1 ($Plc\zeta^{-/-}$) show impaired Ca^{2+} oscillations, polyspermy, and subfertility, suggesting a compensatory or "backup" mechanism to support successful fertilization in the absence of Plcζ1 [10,11]. The egg expresses various isoforms of PLCs, including PLCβs, PLCγs, and PLCδs. While these isoforms have been reported to modulate Ca^{2+} oscillations following fertilization, none can substitute for the essential role of Plcζ in initiating fertilization. For example, overexpression of PLCβ1 in mouse eggs modifies the duration of the sperm-induced Ca^{2+} oscillations' first transient and decreases the frequency of the oscillations [34].

Male infertility rates are rising worldwide [35], impacting the mental health of millions of men by reducing self-esteem and inducing a sense of loss [36]. Mutations in *PLCZ1* are increasingly recognized as a significant genetic cause of male infertility, particularly in individuals presenting with fertilization failure following assisted reproductive techniques such as intracytoplasmic sperm injection (ICSI). Recent studies have reported a prevalence of *PLCZ1* variants as high as 33.6% in men experiencing ICSI failure, underscoring the clinical relevance of this gene in reproductive diagnostics [23]. In clinical practice, *PLCZ1* mutations are typically identified using high-throughput genetic screening methods [13,37,38]. Wholeexome sequencing (WES) is a widely used tool, offering a comprehensive analysis of all coding regions and enabling the identification of known and novel variants. WES is particularly valuable in cases of idiopathic infertility, where there is no prior suspicion of specific genetic defects. Genetic analysis not only supports diagnosis but also guides clinical decision-making in the context of assisted reproductive technologies (ART), including the potential use of oocyte activation protocols or microinjection of wild-type *PLCZ1* mRNA. The microinjection of PLCζ into oocytes can induce Ca^{2+} oscillations needed to start egg activation and early embryo development, which has been demonstrated even in eggs previously injected with PLCζ carrying mutations [39].

4. Materials and Methods

4.1. Full-Atom Molecular Dynamics Simulations and Docking

We used the freely available human PLCζ model predicted by AlphaFold (Alpha-Fold Entry: Q86YW0) [40]. For this, we used as a reference the crystallographic structure of PLC∂ bonded to myo-inositol (PDB ID: 1DJZ) [15] since the active site amino acids are conserved between both enzymes and share a 51% identity. We use the structure of PLCζ in solution rather than at the lipid (membrane)-water interface to decrease the computational cost. Additionally, it has not been reported that PLCζ interacts with membranes. The Ca^{2+} ion was positioned between the residues N171, E200, D202, E249, and the PIP_2 ligand. Then, we used the CHARMM-GUI website to mutate PLCζ and build the systems [41,42]. The histidine 233 was exchanged for leucine (H233L), histidine 398 was converted into a proline (H398P), and the arginine 553 was replaced with a proline (R553P). All systems, wild-type (WT), and the mutants H233L, H398P, and R553P were solvated using the TIP3P water model. The solvated systems were ionized and neutralized with NaCl at a concentration of 150 mM. The water box size was 120 × 119 × 199 Å. Ionized systems were minimized, equilibrated, and run for 1 μs of simulation. MD simulations were performed with AMBER (San Francisco, CA, USA) using the ff19SB force field for proteins and GAFF for PIP_2 ligand [43–45]. The pressure was fixed at 1 atm, and the temperature was kept constant at 310.15 K through the Langevin thermostat with the isobaric-isothermal (NPT) ensemble. Each condition was simulated in three replicates and ran under periodic boundary conditions. After completing the simulations, the subsequent objective was to ascertain the molecular factors associated with the inactivity of PLCζ enzyme mutants. To achieve this, various geometric and energetic stability parameters were computed.

4.2. PIP_2 Mean Square Displacement and Diffusion Coefficient

The Mean Square Displacement (MSD) is a measure of the average squared distance that the molecules travel over time within the binding site. It quantifies the molecule's movement and diffusion, providing insights into its dynamic behavior and interactions with the enzyme during the trajectory. On the other hand, the Diffusion Coefficient (D) of a molecule in the binding site of an enzyme, derived from the MSD, quantifies the rate at which the molecule diffuses within the binding site. It is calculated from the slope of the MSD versus time. MSD and D were computed from the Diffusion Coefficient Tool [46] of VMD v.1.9.3 software (Urbana-Champaign, IL, USA) [47]. This analysis was performed on a PIP_2 ligand placed into the binding site of the PLCζ enzyme. Each calculation involves the quantification of the three axes (x, y, z). The τ value was assigned as the default parameter considering the whole trajectory.

4.3. Binding Free Energy

The protein-ligand affinity energy was computed throughout the entire trajectory at regular intervals of every 10 ns, utilizing the end-point Molecular Mechanics-Generalized

Born Surface Area (MM-GBSA) method of AMBER22 (San Francisco, CA, USA) [48]. This approach accurately estimates the binding free energy between protein and ligand in a molecular system by combining molecular mechanics calculations to represent the protein-ligand complex with a continuum solvent model, the Generalized Born model, to account for solvation effects. By considering both the energetic contributions from the molecular mechanics force field and the solvent interactions, MM-GBSA aims to provide insights into the thermodynamics of binding.

4.4. Protein-Ligand Contacts

Contact types and frequencies were calculated using the GetContacts application (https://getcontacts.github.io/ (accessed on 13 March 2025)). We computed the intermolecular ligand-side chain hydrogen bonds using a donor-acceptor distance < 3.5 Å and an angle of 180°–70°. Likewise, we calculated the ligand-protein salt bridge using a distance cutoff of < 4.0 Å. The van der Waals (vdw) was assessed by employing this equation:

$$|AB| < Rvdw(A) + Rvdw(B) + 0.5$$

where A and B are any non-hydrogen atoms.

4.5. Root Mean Square Fluctuations

The Root Mean Square Fluctuations (RMSF) is a measure of the average deviation of each residue from its average position over time. It quantifies the flexibility and mobility of different parts of the enzyme, indicating which regions are more dynamic and which are more rigid. The enzyme's RMSF profile was calculated using an in-house Tcl script and run in VMD v.1.9.3 software (Urbana-Champaign, IL, USA).

4.6. Ca^{2+} Ion Mean Square Displacement and Diffusion Coefficient

To comprehend the mechanisms involving the Ca^{2+} cofactor and potential chelating effects exhibited by certain mutants, we quantified the ion's MSD and D throughout the simulation, mirroring the methodology employed for the PIP_2 ligand.

4.7. Radial Distribution Function

The Radial Distribution Function (RDF) of water molecules surrounding an ion throughout a simulation measures how the density of water molecules varies as a function of distance from the ion. It provides insights into the spatial organization and structure of the water molecules around the ion. RDF was calculated by averaging the distribution of water molecules at various distances from the ion using the plugin "Radial Pair Distribution Function g(r)" of VMD v.1.9.3 software (Urbana-Champaign, IL, USA).

4.8. Root Mean Square Deviation

The Root Mean Square Deviation (RMSD) of an enzyme through an MD trajectory measures the average deviation of the enzyme's residue positions from a reference structure over time. It quantifies the enzyme's overall conformational changes and structural stability during the simulation. RMSD was calculated along the trajectory using the "RMSD trajectory" tool within VMD v.1.9.3 (Urbana-Champaign, IL, USA). Each domain's backbone of the PLCζ protein was individually delineated as follows: EF-hands (residues 35 to 145), XY (residues 155 to 465), and C2 (residues 466 to 589).

4.9. Intramolecular Hydrogen bonds per Domain

The number of intra-domain hydrogen bonds was determined using the VMD's "Hydrogen bonds" tool. The tool employs default donor-acceptor distance and angle parameters of 3.0 Å and 20°, respectively. The hydrogen bond numbers were quantified in each domain independently.

4.10. Ion Coordination

To evaluate the protein residues and numbers of water molecules coordinating the Ca^{2+} ion, we built an in-house Tcl script that quantifies the protein residues and water molecules within a coordination sphere of 3.5 Å of radius along the simulation time.

4.11. Radius of Gyration

The radius of gyration (Rg) measures the enzyme's compactness and overall shape, providing comprehension of the enzyme's folding, unfolding, and conformational stability during the simulation. The Rg was evaluated with an in-house Tcl script, separately assessing the three PLCζ domains (EF-hands, XY, and C2).

4.12. Statistics

Statistical analysis was performed using GraphPad Prism v.10.4.0 (Boston, MA, USA). The Kolmogorov–Smirnov test was used to determine the data distribution. ANOVA and the Kruskal–Wallis test were used to compare parametric and non-parametric data, respectively. The data were plotted using RStudio software v.2024.09.0+375.

5. Conclusions

Our molecular dynamics simulations provide valuable insights into the structural consequences of PLCζ mutations associated with male infertility. Our findings suggest that mutations such as H233L and H398P disrupt PIP_2 and Ca^{2+} binding and identify residues within the PLCζ structure that stabilize their binding. Understanding the molecular basis of these mutations can aid in developing targeted therapeutic strategies to overcome male infertility. Importantly, these function-abrogating mutations could be the basis for developing non-hormonal contraceptive methods.

Supplementary Materials: The following supporting information can be downloaded at: https://www.mdpi.com/article/10.3390/ijms26104706/s1.

Author Contributions: Conceptualization, F.H. and I.C.; formal analysis, F.H., G.U. and I.A.-D.; investigation, F.H., S.A.-M. and G.U.; writing—original draft preparation, F.H.; writing—review and editing, F.H., S.V., R.A.F., F.D.G.-N., D.B. and I.C.; visualization, F.H., S.A.-M., S.V. and G.U. All authors have read and agreed to the published version of the manuscript.

Funding: This work was funded by ANID FONDECYT Regular 1221308 (ANID-Chile) and Subvención a la Instalación a la Academia SA77210065 (ANID-Chile) projects.

Institutional Review Board Statement: Not applicable.

Informed Consent Statement: Not applicable.

Data Availability Statement: The original contributions presented in this study are included in the article/Supplementary Materials. Further inquiries can be directed to the corresponding author(s).

Acknowledgments: Computational work was supported by the National Laboratory for High Performance Computer (NLHPC, ECM-02), Chile. D.B. would like to acknowledge ANID FONDECYT de Iniciación #11220444 and FOVI230136. S.A-M is recipients of Beca de Doctorado ANID 21252707. We acknowledge BioRender for the schematic images created for this publication (Created with BioRender.com, accessed on 27 July 2022).

Conflicts of Interest: The authors declare no conflicts of interest.

Abbreviations

The following abbreviations are used in this manuscript:

ART	Assisted reproductive technology
DAG	Diacylglycerol
GAFF	The General Amber force field
IP_3	Inositol 1,4,5-triphosphate
IP_3R	Inositol 1,4,5-triphosphate receptor
MD	Molecular dynamics simulations
MM-GBSA	Molecular Mechanics-Generalized Born Surface Area
MSD	Mean square deviation
NPT	Isobaric-isothermal ensemble
PDB	Protein Data Bank
PIP_2	Phosphatidylinositol 4,5-biphosphate
PLCζ	Phospholipase Cζ
PLC∂	Phospholipase C∂
PM	Plasma membrane
RDF	Radial distribution function
Rg	Radius of gyration
RMSD	Root mean square deviation
RMSF	Root mean square fluctuations
vdw	Van der Waals
VMD	Visual molecular dynamics
WT	Wild-type

References

1. Kashir, J.; Jones, C.; Lee, H.C.; Rietdorf, K.; Nikiforaki, D.; Durrans, C.; Ruas, M.; Tee, S.T.; Heindryckx, B.; Galione, A.; et al. Loss of Activity Mutations in Phospholipase C Zeta (PLCζ) Abolishes Calcium Oscillatory Ability of Human Recombinant Protein in Mouse Oocytes. *Hum. Reprod.* **2011**, *26*, 3372–3387. [CrossRef] [PubMed]
2. Kashir, J.; Konstantinidis, M.; Jones, C.; Heindryckx, B.; De Sutter, P.; Parrington, J.; Wells, D.; Coward, K. Characterization of Two Heterozygous Mutations of the Oocyte Activation Factor Phospholipase C Zeta (PLCζ) from an Infertile Man by Use of Minisequencing of Individual Sperm and Expression in Somatic Cells. *Fertil. Steril.* **2012**, *98*, 423–431. [CrossRef] [PubMed]
3. Kashir, J.; Konstantinidis, M.; Jones, C.; Lemmon, B.; Lee, H.C.; Hamer, R.; Heindryckx, B.; Deane, C.M.; De Sutter, P.; Fissore, R.A.; et al. A Maternally Inherited Autosomal Point Mutation in Human Phospholipase C Zeta (PLCζ) Leads to Male Infertility. *Hum. Reprod.* **2012**, *27*, 222–231. [CrossRef] [PubMed]
4. Saunders, C.M.; Larman, M.G.; Parrington, J.; Cox, L.J.; Royse, J.; Blayney, L.M.; Swann, K.; Lai, F.A. PLC Zeta: A Sperm-Specific Trigger of Ca($^{2+}$) Oscillations in Eggs and Embryo Development. *Development* **2002**, *129*, 3533–3544. [CrossRef]
5. Thanassoulas, A.; Swann, K.; Lai, F.A.; Nomikos, M. Sperm factors and egg activation: The Structure and Function Relationship of Sperm PLCZ1. *Reproduction* **2022**, *164*, F1–F8. [CrossRef]
6. Kurokawa, M.; Sato, K.; Fissore, R.A. Mammalian Fertilization: From Sperm Factor to Phospholipase Czeta. *Biol. Cell* **2004**, *96*, 37–45. [CrossRef]
7. Yu, Y.; Nomikos, M.; Theodoridou, M.; Nounesis, G.; Lai, F.A.; Swann, K. PLCζ Causes Ca($^{2+}$) Oscillations in Mouse Eggs by Targeting Intracellular and Not Plasma Membrane PI(4,5)P($_2$). *Mol. Biol. Cell* **2012**, *23*, 371–380. [CrossRef]
8. Miyazaki, S.; Shirakawa, H.; Nakada, K.; Honda, Y. Essential Role of the Inositol 1,4,5-Trisphosphate Receptor/Ca^{2+} Release Channel in Ca^{2+} Waves and Ca^{2+} Oscillations at Fertilization of Mammalian Eggs. *Dev. Biol.* **1993**, *158*, 62–78. [CrossRef]
9. Wakai, T.; Mehregan, A.; Fissore, R.A. Ca^{2+} Signaling and Homeostasis in Mammalian Oocytes and Eggs. *Cold Spring Harb. Perspect. Biol.* **2019**, *11*, a035162. [CrossRef]
10. Hachem, A.; Godwin, J.; Ruas, M.; Lee, H.C.; Ferrer Buitrago, M.; Ardestani, G.; Bassett, A.; Fox, S.; Navarrete, F.; de Sutter, P.; et al. PLCζ Is the Physiological Trigger of the Ca^{2+} Oscillations That Induce Embryogenesis in Mammals but Conception Can Occur in Its Absence. *Development* **2017**, *144*, 2914–2924. [CrossRef]
11. Nozawa, K.; Satouh, Y.; Fujimoto, T.; Oji, A.; Ikawa, M. Sperm-Borne Phospholipase C Zeta-1 Ensures Monospermic Fertilization in Mice. *Sci. Rep.* **2018**, *8*, 1315. [CrossRef] [PubMed]
12. Kashir, J.; Mistry, B.V.; Rajab, M.A.; BuSaleh, L.; Abu-Dawud, R.; Ahmed, H.A.; Alharbi, S.; Nomikos, M.; AlHassan, S.; Coskun, S.; et al. The Mammalian Sperm Factor Phospholipase C Zeta Is Critical for Early Embryo Division and Pregnancy in Humans and Mice. *Hum. Reprod.* **2024**, *39*, 1256–1274. [CrossRef]

13. Heytens, E.; Parrington, J.; Coward, K.; Young, C.; Lambrecht, S.; Yoon, S.-Y.; Fissore, R.A.; Hamer, R.; Deane, C.M.; Ruas, M. et al. Reduced Amounts and Abnormal Forms of Phospholipase C Zeta (PLCzeta) in Spermatozoa from Infertile Men. *Hum Reprod.* **2009**, *24*, 2417–2428. [CrossRef] [PubMed]
14. Yoon, S.-Y.; Jellerette, T.; Salicioni, A.M.; Lee, H.C.; Yoo, M.-S.; Coward, K.; Parrington, J.; Grow, D.; Cibelli, J.B.; Visconti, P.E., et al. Human Sperm Devoid of PLC, Zeta 1 Fail to Induce Ca($^{2+}$) Release and Are Unable to Initiate the First Step of Embryo Development. *J. Clin. Investig.* **2008**, *118*, 3671–3681. [CrossRef]
15. Essen, L.O.; Perisic, O.; Katan, M.; Wu, Y.; Roberts, M.F.; Williams, R.L. Structural Mapping of the Catalytic Mechanism for a Mammalian Phosphoinositide-Specific Phospholipase C. *Biochemistry* **1997**, *36*, 1704–1718. [CrossRef] [PubMed]
16. Kouchi, Z.; Shikano, T.; Nakamura, Y.; Shirakawa, H.; Fukami, K.; Miyazaki, S. The Role of EF-Hand Domains and C2 Domain in Regulation of Enzymatic Activity of Phospholipase Czeta. *J. Biol. Chem.* **2005**, *280*, 21015–21021. [CrossRef]
17. Nomikos, M.; Elgmati, K.; Theodoridou, M.; Calver, B.L.; Nounesis, G.; Swann, K.; Lai, F.A. Phospholipase Cζ Binding to PtdIns(4,5)P_2 Requires the XY-Linker Region. *J. Cell Sci.* **2011**, *124*, 2582–2590. [CrossRef]
18. Nomikos, M.; Mulgrew-Nesbitt, A.; Pallavi, P.; Mihalyne, G.; Zaitseva, I.; Swann, K.; Lai, F.A.; Murray, D.; McLaughlin, S. Binding of Phosphoinositide-Specific Phospholipase C-Zeta (PLC-Zeta) to Phospholipid Membranes: Potential Role of an Unstructured Cluster of Basic Residues. *J. Biol. Chem.* **2007**, *282*, 16644–16653. [CrossRef]
19. Escoffier, J.; Lee, H.C.; Yassine, S.; Zouari, R.; Martinez, G.; Karaouzène, T.; Coutton, C.; Kherraf, Z.-E.; Halouani, L.; Triki, C.; et al. Homozygous Mutation of PLCZ1 Leads to Defective Human Oocyte Activation and Infertility That Is Not Rescued by the WW-Binding Protein PAWP. *Hum. Mol. Genet.* **2016**, *25*, 878–891. [CrossRef]
20. Torra-Massana, M.; Cornet-Bartolomé, D.; Barragán, M.; Durban, M.; Ferrer-Vaquer, A.; Zambelli, F.; Rodriguez, A.; Oliva, R.; Vassena, R. Novel Phospholipase C Zeta 1 Mutations Associated with Fertilization Failures after ICSI. *Hum. Reprod.* **2019**, *34*, 1494–1504. [CrossRef]
21. Nomikos, M.; Stamatiadis, P.; Sanders, J.R.; Beck, K.; Calver, B.L.; Buntwal, L.; Lofty, M.; Sideratou, Z.; Swann, K.; Lai, F.A. Male Infertility-Linked Point Mutation Reveals a Vital Binding Role for the C2 Domain of Sperm PLCζ. *Biochem. J.* **2017**, *474*, 1003–1016. [CrossRef]
22. Yuan, P.; Yang, C.; Ren, Y.; Yan, J.; Nie, Y.; Yan, L.; Qiao, J. A Novel Homozygous Mutation of Phospholipase C Zeta Leading to Defective Human Oocyte Activation and Fertilization Failure. *Hum. Reprod.* **2020**, *35*, 977–985. [CrossRef]
23. Torra-Massana, M.; Rodríguez, A.; Vassena, R. Exonic Genetic Variants Associated with Unexpected Fertilization Failure and Zygotic Arrest after ICSI: A Systematic Review. *Zygote* **2023**, *31*, 316–341. [CrossRef]
24. Pace, C.N.; Fu, H.; Lee Fryar, K.; Landua, J.; Trevino, S.R.; Schell, D.; Thurlkill, R.L.; Imura, S.; Scholtz, J.M.; Gajiwala, K.; et al. Contribution of Hydrogen bonds to Protein Stability. *Protein Sci.* **2014**, *23*, 652–661. [CrossRef]
25. World Health Organization. *Infertility Prevalence Estimates, 1990–2021*; World Health Organization: Geneva, Switzerland, 2023.
26. Nomikos, M.; Blayney, L.M.; Larman, M.G.; Campbell, K.; Rossbach, A.; Saunders, C.M.; Swann, K.; Lai, F.A. Role of Phospholipase C-Zeta Domains in Ca^{2+}-Dependent Phosphatidylinositol 4,5-Bisphosphate Hydrolysis and Cytoplasmic Ca^{2+} Oscillations. *J. Biol. Chem.* **2005**, *280*, 31011–31018. [CrossRef]
27. Kouchi, Z.; Fukami, K.; Shikano, T.; Oda, S.; Nakamura, Y.; Takenawa, T.; Miyazaki, S. Recombinant Phospholipase Czeta Has High Ca^{2+} Sensitivity and Induces Ca^{2+} Oscillations in Mouse Eggs. *J. Biol. Chem.* **2004**, *279*, 10408–10412. [CrossRef]
28. Ellis, M.V.; Sally, U.; Katan, M. Mutations within a Highly Conserved Sequence Present in the X Region of Phosphoinositide-Specific Phospholipase C-Delta 1. *Biochem. J.* **1995**, *307 Pt 1*, 69–75. [CrossRef]
29. Ellis, M.V.; James, S.R.; Perisic, O.; Downes, C.P.; Williams, R.L.; Katan, M. Catalytic Domain of Phosphoinositide-Specific Phospholipase C (PLC). Mutational Analysis of Residues within the Active Site and Hydrophobic Ridge of Plcdelta1. *J. Biol. Chem.* **1998**, *273*, 11650–11659. [CrossRef] [PubMed]
30. Hinostroza, F.; Neely, A.; Araya-Duran, I.; Marabolí, V.; Canan, J.; Rojas, M.; Aguayo, D.; Latorre, R.; González-Nilo, F.D.; Cárdenas, A.M. Dynamin-2 R465W Mutation Induces Long Range Perturbation in Highly Ordered Oligomeric Structures. *Sci. Rep.* **2020**, *10*, 18151. [CrossRef] [PubMed]
31. Das, J.K.; Thakuri, B.; MohanKumar, K.; Roy, S.; Sljoka, A.; Sun, G.-Q.; Chakraborty, A. Mutation-Induced Long-Range Allosteric Interactions in the Spike Protein Determine the Infectivity of SARS-CoV-2 Emerging Variants. *ACS Omega* **2021**, *6*, 31305–31320. [CrossRef] [PubMed]
32. Nomikos, M. Novel Signalling Mechanism and Clinical Applications of Sperm-Specific PLCζ. *Biochem. Soc. Trans.* **2015**, *43*, 371–376. [CrossRef]
33. Theodoridou, M.; Nomikos, M.; Parthimos, D.; Gonzalez-Garcia, J.R.; Elgmati, K.; Calver, B.L.; Sideratou, Z.; Nounesis, G.; Swann, K.; Lai, F.A. Chimeras of Sperm PLCζ Reveal Disparate Protein Domain Functions in the Generation of Intracellular Ca^{2+} Oscillations in Mammalian Eggs at Fertilization. *Mol. Hum. Reprod.* **2013**, *19*, 852–864. [CrossRef]
34. Igarashi, H.; Knott, J.G.; Schultz, R.M.; Williams, C.J. Alterations of PLCβ1 in Mouse Eggs Change Calcium Oscillatory Behavior Following Fertilization. *Dev. Biol.* **2007**, *312*, 321–330. [CrossRef] [PubMed]

35. Skakkebaek, N.E.; Rajpert-De Meyts, E.; Buck Louis, G.M.; Toppari, J.; Andersson, A.-M.; Eisenberg, M.L.; Jensen, T.K.; Jørgensen, N.; Swan, S.H.; Sapra, K.J.; et al. Male Reproductive Disorders and Fertility Trends: Influences of Environment and Genetic Susceptibility. *Physiol. Rev.* **2016**, *96*, 55–97. [CrossRef]
36. Nachtigall, R.D.; Becker, G.; Wozny, M. The Effects of Gender-Specific Diagnosis on Men's and Women's Response to Infertility. *Fertil. Steril.* **1992**, *57*, 113–121. [CrossRef]
37. Mu, J.; Zhang, Z.; Wu, L.; Fu, J.; Chen, B.; Yan, Z.; Li, B.; Zhou, Z.; Wang, W.; Zhao, L.; et al. The Identification of Novel Mutations in PLCZ1 Responsible for Human Fertilization Failure and a Therapeutic Intervention by Artificial Oocyte Activation. *Mol. Hum. Reprod.* **2020**, *26*, 80–87. [CrossRef]
38. Yan, Z.; Fan, Y.; Wang, F.; Yan, Z.; Li, M.; Ouyang, J.; Wu, L.; Yin, M.; Zhao, J.; Kuang, Y.; et al. Novel Mutations in PLCZ1 Cause Male Infertility Due to Fertilization Failure or Poor Fertilization. *Hum. Reprod.* **2020**, *35*, 472–481. [CrossRef]
39. Nomikos, M.; Yu, Y.; Elgmati, K.; Theodoridou, M.; Campbell, K.; Vassilakopoulou, V.; Zikos, C.; Livaniou, E.; Amso, N.; Nounesis, G.; et al. Phospholipase Cζ Rescues Failed Oocyte Activation in a Prototype of Male Factor Infertility. *Fertil. Steril.* **2013**, *99*, 76–85. [CrossRef]
40. Jumper, J.; Evans, R.; Pritzel, A.; Green, T.; Figurnov, M.; Ronneberger, O.; Tunyasuvunakool, K.; Bates, R.; Žídek, A.; Potapenko, A.; et al. Highly Accurate Protein Structure Prediction with AlphaFold. *Nature* **2021**, *596*, 583–589. [CrossRef]
41. Jo, S.; Kim, T.; Iyer, V.G.; Im, W. CHARMM-GUI: A Web-Based Graphical User Interface for CHARMM. *J. Comput. Chem.* **2008**, *29*, 1859–1865. [CrossRef]
42. Lee, J.; Cheng, X.; Swails, J.M.; Yeom, M.S.; Eastman, P.K.; Lemkul, J.A.; Wei, S.; Buckner, J.; Jeong, J.C.; Qi, Y.; et al. CHARMM-GUI Input Generator for NAMD, GROMACS, AMBER, OpenMM, and CHARMM/OpenMM Simulations Using the CHARMM36 Additive Force Field. *J. Chem. Theory Comput.* **2016**, *12*, 405–413. [CrossRef]
43. Brooks, B.R.; Brooks, C.L.; Mackerell, A.D.; Nilsson, L.; Petrella, R.J.; Roux, B.; Won, Y.; Archontis, G.; Bartels, C.; Boresch, S.; et al. CHARMM: The Biomolecular Simulation Program. *J. Comput. Chem.* **2009**, *30*, 1545–1614. [CrossRef]
44. Gao, Y.; Lee, J.; Smith, I.P.S.; Lee, H.; Kim, S.; Qi, Y.; Klauda, J.B.; Widmalm, G.; Khalid, S.; Im, W. CHARMM-GUI Supports Hydrogen Mass Repartitioning and Different Protonation States of Phosphates in Lipopolysaccharides. *J. Chem. Inf. Model.* **2021**, *61*, 831–839. [CrossRef]
45. Lee, J.; Hitzenberger, M.; Rieger, M.; Kern, N.R.; Zacharias, M.; Im, W. CHARMM-GUI Supports the Amber Force Fields. *J. Chem. Phys.* **2020**, *153*, 035103. [CrossRef]
46. Giorgino, T. Computing Diffusion Coefficients in Macromolecular Simulations: The Diffusion Coefficient Tool for VMD. *J. Open Source Softw.* **2019**, *4*, 1698. [CrossRef]
47. Humphrey, W.; Dalke, A.; Schulten, K. VMD: Visual Molecular Dynamics. *J. Mol. Graph.* **1996**, *14*, 33–38. [CrossRef]
48. Case, D.A.; Aktulga, H.M.; Belfon, K.; Ben-Shalom, I.Y.; Berryman, J.T.; Brozell, S.R.; Cerutti, D.S.; Cheatham, T.E.; Cisneros, G.A.; Cruzeiro, V.W.D.; et al. *Amber 2022*; University of California: San Francisco, CA, USA, 2022.

International Journal of
Molecular Sciences

MDPI

Article

Evolutionary Origins and Functional Diversification of 2′-O-Methyltransferases: Insights from Phylogenetic and Structural Analysis

Sai-Nan Wang, Xiao-Xia Liu, Ling-Jie Lei, Qiang Wang, Zhu-Qing Shao * and Yang Liu *

School of Life Sciences, Nanjing University, Nanjing 210023, China; dg21300039@smail.nju.edu.cn (S.-N.W.); 502022300022@smail.nju.edu.cn (X.-X.L.); 502022300013@smail.nju.edu.cn (L.-J.L.); wangq@nju.edu.cn (Q.W.)
* Correspondence: zhuqingshao@nju.edu.cn (Z.-Q.S.); 2111604@nju.edu.cn (Y.L.)

Abstract: Ribose 2′-O-methylation (Nm), a key RNA modification, is catalyzed by diverse 2′-O-methyltransferases (2′-O-MTases), yet the evolutionary trajectories of these enzymes remain poorly studied. Here, with a comprehensive collection of functionally validated 2′-O-MTases, we classified them into 11 families based on the distinct methyltransferase (MTase) domains. Homology searches across 198 species identified 6746 proteins, revealing the widespread distribution of 2′-O-MTases across the Tree of Life. Eight MTase domains (e.g., FtsJ, SpoU-methylase) existed both in eukaryotes and prokaryotes, indicating their ancient origin in the Last Universal Common Ancestor (LUCA). In contrast, the AdoMet-MTase, TRM13, and Trm56 domains are lineage-specific. Copy number expansion of most 2′-O-MTase families occurred as life evolved from prokaryotes to eukaryotes, where they might engage in more complex regulation of cell differentiation and development. Domain composition, Ka/Ks ratio, and domain structural analyses showed that purifying selection conserved catalytic domains across most families, despite the frequent integration of auxiliary domains. Notably, the FtsJ family diverged into three deeply separated lineages via remodeling the catalytic pocket, with each lineage specializing in the methylation of mRNA caps, rRNA, or tRNA. These findings illuminate the evolutionary trajectory of 2′-O-MTases, highlighting their ancient multiple origins and functional diversification.

Keywords: RNA modification; methylation; 2′-O-methyltransferase; phylogenetic analysis; structural evolution; substrate adaptability

1. Introduction

RNA modifications, including N6-methyladenosine (m^6A), 5-methylcytosine (m^5C), 2′-O-methylation (Nm), N1-methyladenosine (m^1A), N7-methylguanosine (m^7G), N4-acetylcytosine (ac^4C), and pseudouridine (Ψ), function as pivotal epigenetic regulators, orchestrating both co-transcriptional and post-transcriptional gene expression processes [1]. Among these, Nm is a highly abundant modification, occurs at the ribose sugar of nucleotides, and is found across multiple RNA types, including rRNA, tRNA, non-coding RNA, and mRNA [2].

In rRNA, Nm modifications stabilize the ribosome structure for translational accuracy [3], while in tRNA, they enhance tRNA thermal stability and decoding efficiency [4]. Small RNAs such as animal piRNAs and plant miRNAs acquire 3′-terminal Nm modifications to resist enzymatic degradation, ensuring their longevity [5]. In mRNA, 5′-cap Nm modifications not only stabilize the transcript but also help distinguish host transcripts from viral RNA, preventing immune sensor recognition [6]. Emerging evidence suggests that internal Nm in mRNA may influence translation efficiency and splicing, though the mechanisms behind this remain unclear [7,8].

Nm modifications are catalyzed by two distinct enzymatic systems: C/D-box small nucleolar ribonucleoprotein (snoRNP) complexes and guide RNA-independent 2′-O-MTases [2]. The Fibrillarin-containing C/D-box snoRNP complex catalyzes rRNA 2′-O-methylation in archaea and eukaryotes via a base-pairing process driven by guide RNA. To transfer methyl groups from S-adenosylmethionine (SAM) to the ribose 2′-OH at specified sites (e.g., 18S/28S rRNA in humans), Fibrillarin, the 2′-O-MTase, is positioned by the conserved C/D-box motifs and antisense element complementary to the target rRNA sequence found in the snoRNA component [3,9]. Guide RNA-independent methyltransferases, on the other hand, are able to directly identify the structural or sequence patterns of their substrates. For instance, TRM7 in humans binds the L-shaped tertiary structure of tRNA to create a heterodimer with Trm732 to methylate sites 32 and 34 in the anticodon loop, a process crucial to the precision of decoding [10]. Similarly to this, HEN1 catalyzes 3′-terminal 2′-O-methylation to prevent exonucleolytic degradation by recognizing the 2-nt 3′ overhang of plant miRNAs and animal piRNAs via a flexible loop that contains the FXPP motif (N-terminal to the MTase domain) [5]. The 5′-cap structure of mRNA in eukaryotes is subject to successive 2′-O-methylation: CMTR1 methylates the first transcribed nucleotide's ribose 2′-OH, whereas CMTR2 alters the second nucleotide [11]. Since unmethylated cap0 structures are recognized by innate immune sensors (such RIG-I/MDA5) as pathogen-associated molecular patterns, this hierarchical methylation is essential for differentiating between viral and self-RNA [12,13]. Surprisingly, coronaviruses like SARS-CoV-2 imitate host CMTR1 methylation by taking advantage of a 2′-O-MTase complex (nsp16/nsp10) [14–16]. While nsp10 stabilizes the RNA-binding interface, the viral nsp16, which is structurally similar to CMTR1, methylates the first nucleotide's 2′-OH utilizing SAM as a methyl donor and binds the cap0 structure via its K-D-K-E catalytic tetrad. Viral RNA can avoid host immune monitoring thanks to this mimicry, which is an example of an evolutionary arms race focused on 2′-O-MTase activity.

Fibrillarin is widespread across the Tree of Life, with mammalian Fibrillarin duplicating into two subgroups following the *Eutheria split*, suggesting potential functional diversification. Functional specialization resulting from this duplication has been demonstrated in *S. mediterranea* [17,18]. Phylogenetic analysis of the plant HEN1 protein family revealed several highly conserved motifs and intensive gene duplication in the evolution of the HEN1 family in flowering plants, which may be associated with whole-genome duplication events [19]. Similarly, evolutionary analyses suggest that at least one SPOUT member involved in 2′-O-methylation was present in the LUCA [20]. In addition to Fibrillarin, HEN1, and SPOUT, several other 2′-O-MTases have been found to participate in RNA 2′-O-methylation; these include TRM7 [21], FtsJ3 [22], TRM13 [23], and RSMI [24]; however, their evolutionary background and origins are still unclear. Understanding the development of Nm changes in various life domains requires a systematic evolutionary framework that includes a variety of 2′-O-MTases from the Tree of Life.

We report a phylogenetic analysis of 2′-O-MTases across 198 genomes from bacteria, archaea, and eukaryotes in this work, which involved a thorough collection and domain annotation of functionally proven 2′-O-MTases. We were able to determine their ancient origin and unique evolutionary patterns by combining domain architecture, selection pressure (Ka/Ks), and structural comparisons. Furthermore, we discovered that the FtsJ family was divided into three subclasses, each with unique substrate specificities and evolutionary-shaped structural characteristics. Our results present a thorough analysis of the history of 2′-O-MTases over a period of four billion years, offering fresh perspectives on their genesis and diversity.

2. Results

2.1. Literature-Based Collection of 2′-O-MTases Reveals 11 Distinct MTase Domains for Nm Catalyzation

To identify catalytic domains of 2′-O-MTases, we conducted a thorough survey of functionally known 2′-O-MTases reported in the literature. A total of 36 experimentally verified 2′-O-MTases were retrieved (Table S1), which represent 27 non-orthologous functional genes (Table S1). Hmmscan analysis revealed 11 distinct MTase domains from the protein sequences of these enzymes, including FtsJ (PF01728.24.hmm), Fibrillarin (PF01269.22.hmm), TP-methylase (PF00590.25.hmm), SpoU-methylase (PF00588.24.hmm), AdoMet-MTase (PF07757.18.hmm), TRM13 (PF05206.19.hmm), Trm56 (PF01994.21.hmm), Methyltransf-11 (PF08241.17.hmm), Methyltransf-12 (PF08242.17.hmm), Methyltransf-23 (PF13489.11.hmm), and Methyltransf-31 (PF13847.11.hmm), with domain names corresponding to the official Pfam annotations (Figure 1).

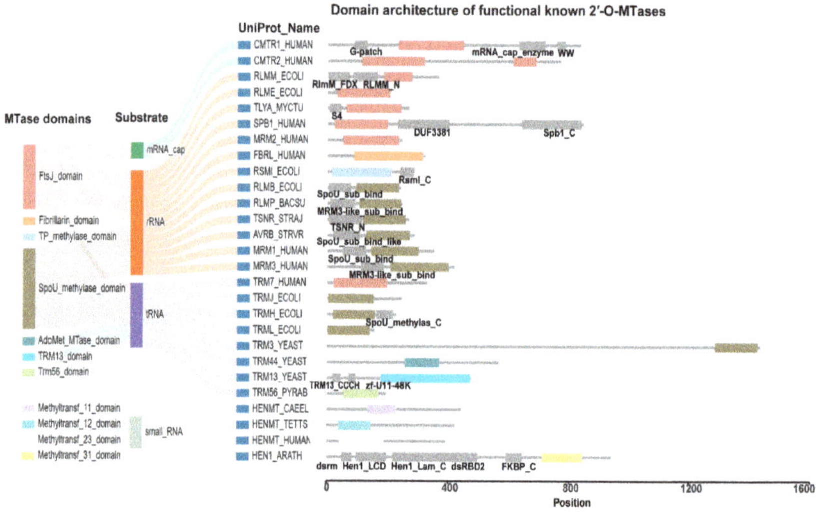

Figure 1. Domain annotation and classification of functionally known 2′-O-MTases. A total of 27 functionally characterized 2′-O-methyltransferases (2′-O-MTases) with distinct substrate specificities were retrieved from the literature. The UniProt Protein names of all proteins are provided, and full details including domain positions are listed in Supplementary Table S1. Based on domain composition, these enzymes were classified into 11 families, each sharing a conserved methyltransferase (MTase) domain. In the domain architecture diagram, colored domains correspond to the MTase domains shown in the first part of the Sankey diagram and are not duplicated. Gray segments represent other functional domains outside of the methylation-related MTase domain and are labeled directly in the figure. Domain annotations were obtained using HMMER3 with Pfam models, and domain names follow the InterPro convention.

Among the 11 defined 2′-O-MTase families, most of them modify specific RNA substrates. For instance, TP-methylase modifies small rRNA [24], while TRM13 targets tRNA [23] (Figure 1). However, 2′-O-MTases with two domains exhibit broader specificity. For example, FtsJ family members catalyze Nm modifications on mRNA caps (CMTR1/2), rRNA (RLMM/RLME/TLYA/SPB1/MRM2), and tRNA (TRM7). Similarly, SpoU-methylase modifies both rRNA and tRNA (Figure 1).

Interestingly, an extra RNA-binding domain called SpoU-sub-bind (like) is typically present in the SpoU-methylase family members that exclusively catalyze rRNA Nm alterations (Figure 1). On the other hand, SpoU-methylase family members that modify tRNA usually have a simpler domain design that just includes the SpoU-methylase domain (Figure 1). The DHU loops 42 and 43 (TRMH/TRM3) and the anticodon loops 44 and 45 (TRMJ/TRML) are examples of fixed loops where the target tRNA modification sites for SpoU-methylase proteins are frequently found [25–28]. However, the FtsJ family did not exhibit any other substrate-related domains. Surprisingly, the various proteins in this family can identify their particular substrates with just the FtsJ domain—no additional RNA-binding domains are required. According to this, the FtsJ domain could have developed an innate method for identifying a broad range of RNA targets, allowing for significant substrate diversity and adaptation across various RNA types.

2.2. Phylogenetic Distribution of the 11 MTase Domain-Containing Proteins Suggests Ancient and Life Kingdom-Specific Origins of 2′-O-MTases

For 11 MTase domains identified from 2′-O-MTases, we analyzed their distribution across 198 species from the Tree of Life, including 19 bacterial groups (44 species), 14 archaeal groups (32 species), 10 fungal groups (29 species), 23 animal groups (60 species), and 9 plant groups (33 species). The results show that proteins containing the FtsJ, SpoU-methylase, TP-methylase, and Methyltransf-11, -12, -23, and -31 domains are widely distributed across eukaryotes and prokaryotes, suggesting that these 2′-O-MTase families originate from the LUCA. Notably, Fibrillarin is conserved across eukaryotes and archaea. The TRM13 family is absent from bacteria and archaea, indicating its origin in the common ancestor of eukaryotes. The AdoMet-MTase family is present in fungi and animals, and the Trm56 family is specifically found in archaea (Figure 2A).

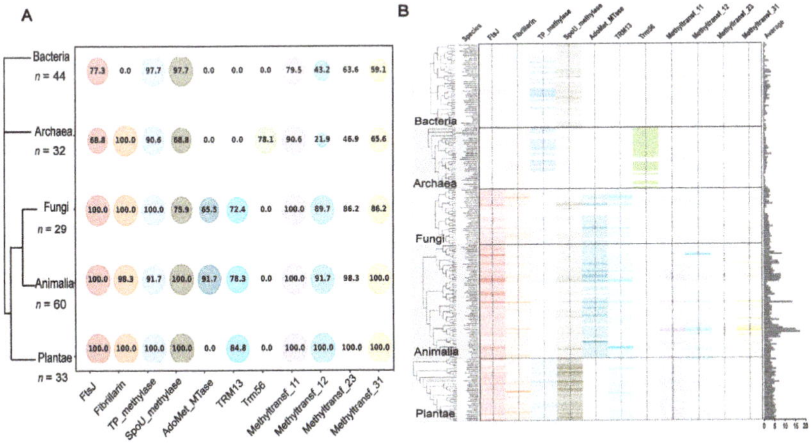

Figure 2. Phylogenetic distribution of the 11 2′-O-MTase families across the Tree of Life. (**A**). The proportion of the investigated species containing each of the 2′-O-MTase family genes across the bacteria, archaea, fungi, animals, and plants. Color-coded bubbles represent 11 families, with size indicating the proportion of species containing each family (largest circle = 100%). (**B**). Abundance of different 2′-O-MTases in each species. Colors intensity indicating relative quantity. Gray bars show the average number of 2′-O-MTases per species.

Generally, the total 2′-O-MTase gene copy number increased during the prokaryote-to-eukaryote transition, reflecting the growing complexity of RNA modifications (Supplementary Figure S1). Among the eight families present in both prokaryotes and eukaryotes, FtsJ, SpoU-methylase, and Methyltransf-11, -12, -23, and -31 families show an overall increase, the Fibrillarin family is relative stable, and TP-methylase is enriched in prokaryotes and plants (Figure 2B, Table S3).

2.3. *Diverse Domain Architectures of 2′-O-MTases*

To explore functional diversification among 2′-O-MTases, we analyzed domain architectures across 11 families. AdoMet-MTase, Fibrillarin, and Trm56 exhibit the minimal domain complexity, with most proteins maintaining a single-domain (Figure 3A–K, Table S4). In contrast, the FtsJ family frequently combines the core MTase domain with the Spb1-C and DUF3381 domains (Figure 3A), which are essential for ribosome assembly [29]. SpoU-methylase often fuses the SpoU-sub-bind domain, likely enhancing rRNA binding specificity (Figure 3D). The four Methyltransf families exhibit complex domain architectures, potentially linked to diverse regulatory mechanisms (Figure 3H–K, Table S4).

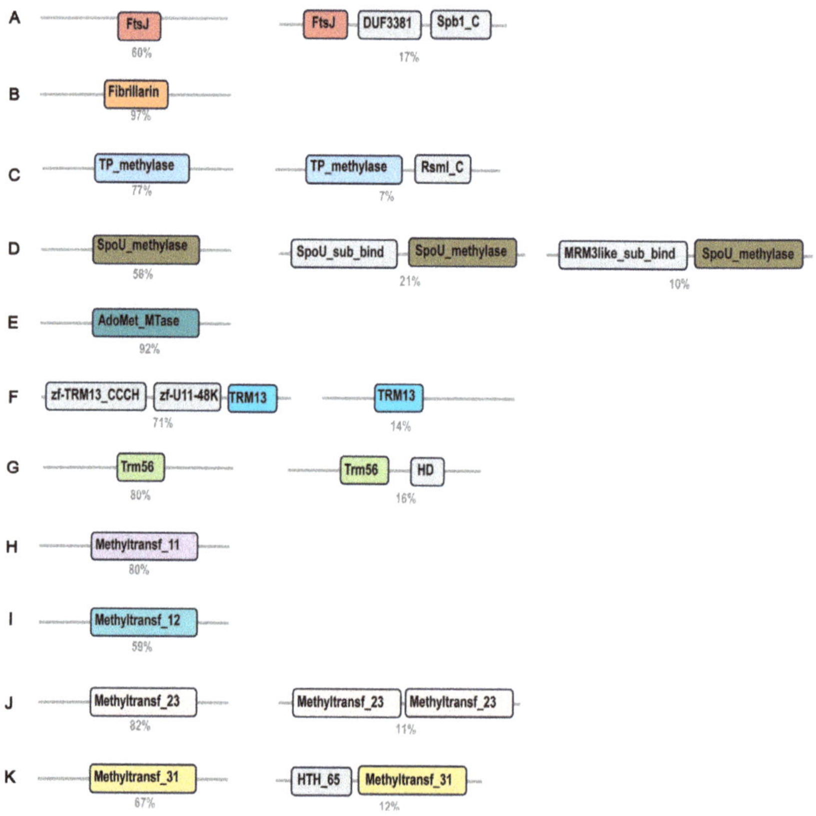

Figure 3. This figure illustrates the domain architectures of each 2′-O-MTase family (**A–K**). Each subfigure shows the domain architectures for a specific 2′-O-MTase family. Colored domains represent the 11 MTase domains that we studied, while gray domains represent other domains. Only the combinations that account for 10% or more of the total proteins in each family are shown.

2.4. *Evolutionary Features of Sequence and Structure of 2′-O-MTases*

An integrated analysis comprising sequence similarity analysis (Figure 4A, Table S5), selective pressure analysis (Figure 4B, Table S6), and structural similarity analysis (Figure 4C, Table S7) was carried out to investigate the evolutionary trends of 2′-O-MTases. In eukaryotes, all 10 families showed Ka/Ks ratios consistently below 1 (Figure 4B), despite differing levels of sequence divergence (Figure 4A), suggesting a general purifying selection. Among the 11 2′-O-MTases, Fibrillarin exhibited exceptional sequence conservation (an average of 57.95% sequence similarity cross different life kingdoms and 75.92% among eukaryotes), ultralow Ka/Ks ratios (0.0259–0.0546), and high structural conservation (average TM score = 0.85), indicating strong purifying selection on the catalytic core, consistent with its essential role, while gene duplication likely led to regulatory diversification (see Discussion). The TRM13 and Trm56 families, on the other hand, showed relaxed purifying selection (Ka/Ks: 0.21–0.34) and moderate sequence plasticity (11.1–41.9% similarity).

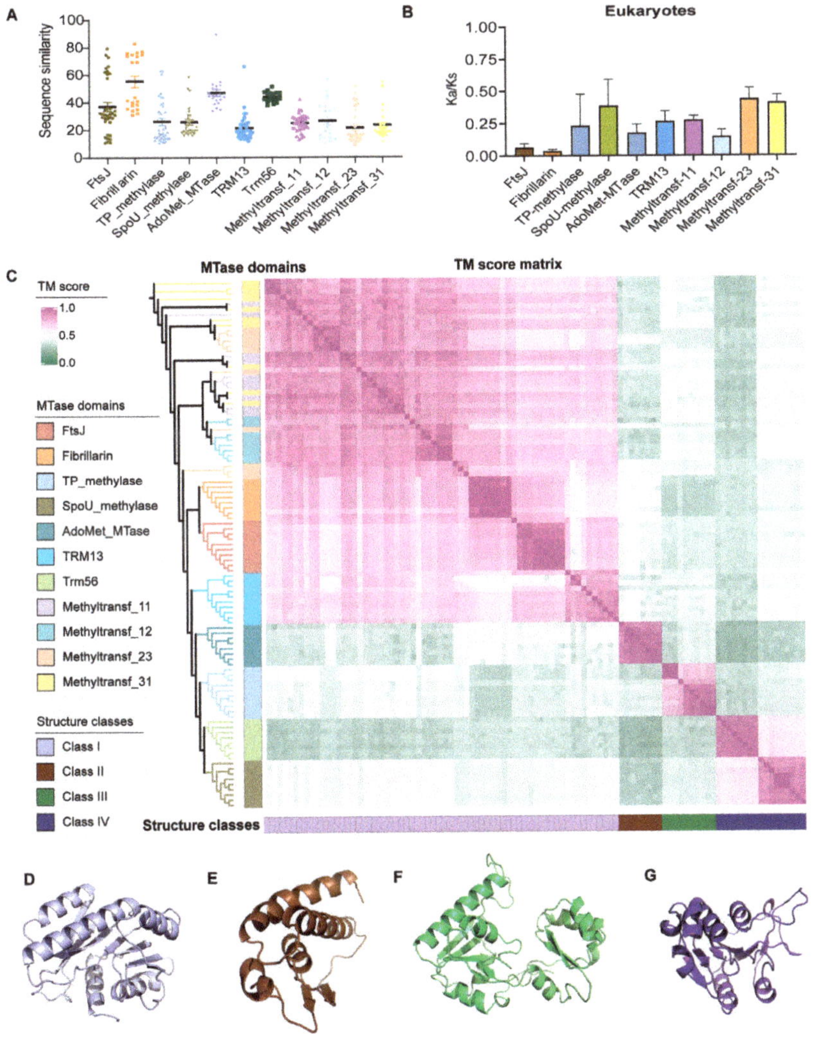

Figure 4. Sequence similarity, selective pressure, and structural similarity of 2′-O-MTases. (**A**). Sequence similarity of 11 2′-O-MTases. Colored scatter points represent pairwise similarity values across families. (**B**). Selective pressure of each 2′-O-MTase family in eukaryotes. (**C**). Structural similarity and clustering of 11 2′-O-MTase families across 10 species. Matrix plot shows pairwise TM-scores, with color intensity indicating structural similarity. (**D**). Representative structure of Class 1 (NP_497843.1_Caenorhabditis-elegans_FtsJ). (**E**). Representative structure of Class 2 (NP_741176.2_Caenorhabditis-elegans_AdoMet_MTase). (**F**). Representative structure of Class 3 (WP_031175390.1_Streptomyces-albus_TP_methylase). (**G**). Representative structure of Class 4 (NP_495375.3_Caenorhabditis-elegans_SpoU_methylase).

The hierarchical clustering resolved 2′-O-MTases into four conserved structural classes, each with a TM score average of 0.74. These are Class I: Methyltransf-11, -12, -23, and -31, FtsJ, Fibrillarin, and TRM13; Class II: AdoMet-MTase; Class III: TP-methylase; and Class IV: SpoU-methylase and Trm56 (Figure 4C). Class I enzymes typically feature a seven-stranded β-sheet surrounded by α-helices, forming an open αβα sandwich structure. Class II enzymes usually have a two-stranded β-sheet along with three α-helices. Class III enzymes possess a bilobal structure. Class IV enzymes typically possess a six-stranded parallel β-sheet and seven α-helices, with the C terminus forming a unique "knot" structure (Figure 4D,E). Notably, within Class 1, we observed striking spatial structural similarity across distinct families. For example, subclades of Methyltransf-11, -12, -23, and -31 exhibited intermixed distributions with an average TM-score of 0.82, while FtsJ and Fibrillarin shared high structural similarity (average: 0.84). Given that structural evolution is slower than sequence divergence [30], these findings suggest that these families may share a common evolutionary origin, despite their sequence-level divergence.

Furthermore, we computed the correlation between structural divergence (1-TM score) and sequence evolutionary distance across 2′-O-MTase families in order to evaluate the link between sequence evolution and structural conservation. Structural changes trail sequence divergence, as evidenced by the strong positive correlation (slope < 1; Figure 5A–K). The inverse slope (1/slope), which is a measure of structural tolerance, identified conservation tactics unique to each family: Fibrillarin maintained high structural similarity (structure similarity score > 0.8) across kingdoms, even with moderate sequence divergence (~58% sequence similarity), demonstrating strong structural robustness (1/slope = 23.55; Figure 5L). The Trm56 (1/slope = 9.70) and TRM13 (1/slope = 12.07) families, on the other hand, showed constrained plasticity, allowing for substantial sequence variation (11.1–41.9% similarity) while maintaining structural integrity.

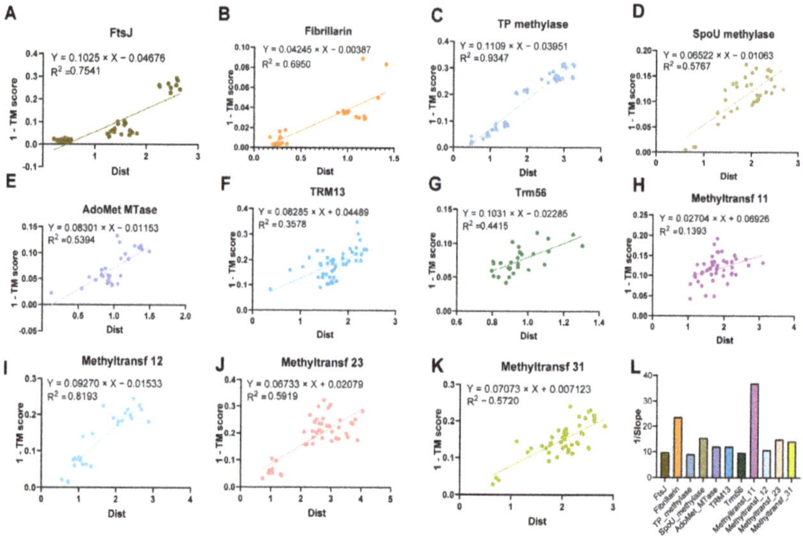

Figure 5. Structural sensitivity of 2′-O-MTases. (**A–K**). Structural distance versus evolutionary divergence in 11 2′-O-MTase families. Scatter plots showing the relationship between structural distance (1-TM score) and evolutionary divergence (amino acid substitutions per site) for (**A**) FtsJ, (**B**) Fibrillarin, (**C**) TP-methylase, (**D**) SpoU-methylase, (**E**) AdoMet-MTase, (**F**) TRM13, (**G**) Trm56, (**H**) Methyltransf-11, (**I**) Methyltransf-12, (**J**) Methyltransf-23, and (**K**) Methyltransf-31. Correlation equations and R^2 values are shown for each family. (**L**). Bar plot of 1/slope values for each family.

2.5. Ancient Divergence and Substrate-Adaptive Evolution of the FtsJ Protein Family

As previously mentioned in Figure 1, despite having identical individual domain designs, the FtsJ family proteins show extraordinary substrate variety, with several members performing modifications on rRNA, tRNA, and mRNA-cap. Phylogenetic analysis of 679 FtsJ-domain-containing proteins from 198 species revealed three deeply separated lineages (Figure 6A). It is interesting to note that FtsJ proteins with varying substrate specificities for rRNA, tRNA, and mRNA-cap were found in three separate lineages, indicating that lineage divergence may have contributed to the development of substrate specificity in FtsJ proteins. This pattern suggests that FtsJ proteins may have been specialized for rRNA, tRNA, and mRNA cap modification as a result of each lineage's adaptation to certain substrate types.

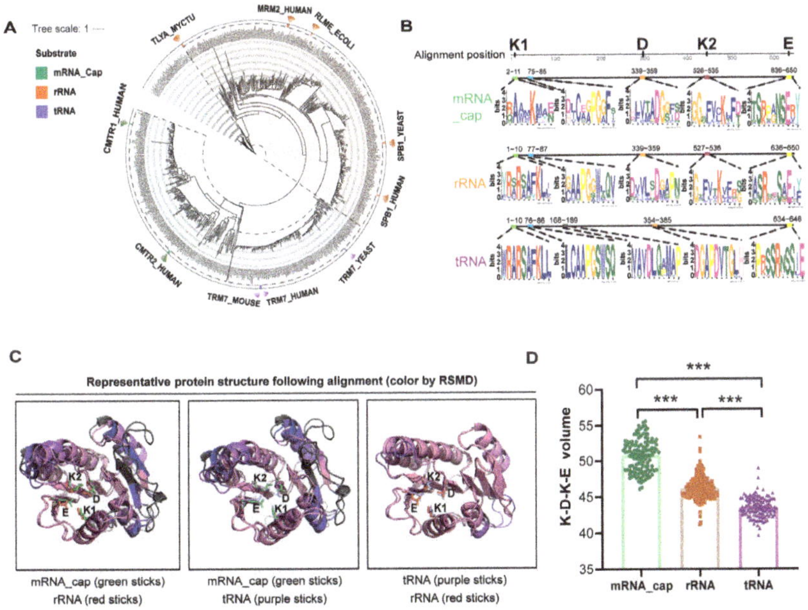

Figure 6. Subclass-specific features in FtsJ family. (**A**). Phylogenetic analysis of FtsJ subclasses. Maximum likelihood tree shows the evolutionary relationships among mRNA-cap (green, $n = 191$), rRNA (red, $n = 316$), and tRNA (purple, $n = 139$) subclasses. (**B**). Motif analysis of FtsJ subclasses. Sequence logos depict conserved motifs, with the catalytic tetrad (KDKE) highlighted. (**C**). PyMOL-generated alignments of representative proteins: mRNA-cap (Q8N1G2-CMTR1-HUMAN), rRNA (Q8IY81-SPB1-HUMAN), and tRNA (Q9UET6-TRM7-HUMAN). Structures are colored according to per-residue RMSD: pink indicates regions with low RMSD (high structural similarity), blue denotes regions with high RMSD (greater structural divergence), and gray highlights structural elements present in only one of the aligned proteins. The KDKE catalytic tetrad residues are displayed in stick representation. (**D**). Catalytic pocket volume across FtsJ subclasses. Bar plot compares KDKE pocket volumes for mRNA-cap, rRNA, and tRNA subclasses. Error bars represent the SEM for each subclass. *** indicates $p < 0.001$.

Motif analysis showed that all FtsJ subclasses retain the essential KDKE catalytic core but exhibit subclass-specific features (Figure 6B). As previously reported, K1 stabilizes the RNA phosphate group, D binds SAM, K2 facilitates nucleophilic attack, and E coordinates K1 and K2 [31–34]. Despite KDKE conservation, each subclass exhibits unique sequence characteristics. For instance, the mRNA-cap subclass is characterized by a distinctive residue composition, with K1 frequently forming the RAAMK motif. On the other hand, the RSAFK motif is commonly formed by both the rRNA and tRNA subclasses. In contrast to the other subclasses, the tRNA subclass notably exhibits greater sequence conservation, indicating a potentially more stringent structural requirement for its target changes. These varied evolutionary paths highlight the FtsJ domain's structural flexibility, which permits functional adaptability to a variety of RNA substrates while maintaining basic catalytic activity.

Structural prediction and alignment revealed high intra-subclass consistency but notable inter-subclass divergence, particularly in the N-terminal regions of mRNA-cap modifiers outside the conserved KDKE catalytic core (Figure 6C). These variations likely enhance substrate recognition, driving functional diversification. Quantitative analysis of the KDKE catalytic tetrad showed subclass-specific features: mRNA-cap modifiers exhibited the largest catalytic cage volume (Figure 6D), accompanied by an expanded K1-D spacing (Figure 7A), whereas the rRNA/tRNA subclasses retained compact configurations. These structural differences may reflect evolutionary adaptations to substrate size and complexity—the enlarged cage of mRNA-cap modifiers accommodates the bulky 5′ cap, while the narrower pocket in tRNA subclasses matches the anticodon loop (Figure 7A–G).

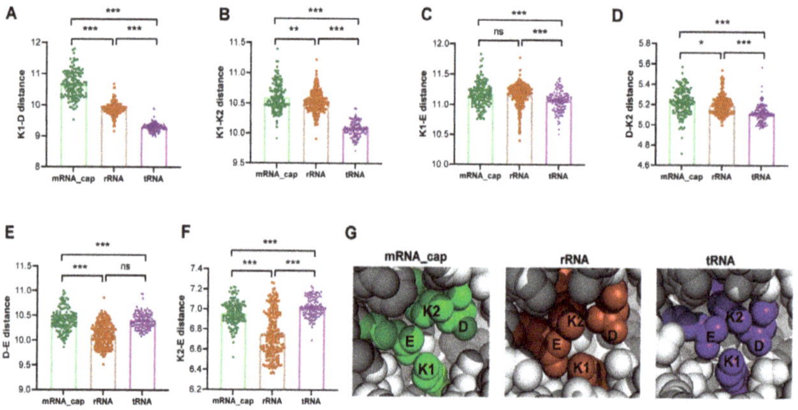

Figure 7. Subclass-specific structural variations in the KDKE catalytic tetrad in the FtsJ family. A-E. Distance analysis of KDKE catalytic tetrad residues. (**A**) K1-D, (**B**) K1-K2, (**C**) K1-E, (**D**) D-K2, (**E**) D-E, and (**F**) K2-E distances across FtsJ subclasses: mRNA-cap (green), rRNA (red), and tRNA (purple). Distances were calculated based on Cα atoms. (**G**) Structural alignment of FtsJ subtypes. PyMOL-generated alignments of representative proteins: mRNA-cap (Q8N1G2-CMTR1-HUMAN, green), rRNA (Q8IY81-SPB1-HUMAN, red), and tRNA (Q9UET6-TRM7-HUMAN, purple), shown as spheres. The KDKE tetrad is highlighted in each structure. Error bars represent the SEM for each subclass. * indicates $p < 0.05$, ** indicates $p < 0.01$, *** indicates $p < 0.001$, and ns indicates not significant.

3. Discussion

2′-O-methylation, catalyzed by 2′-O-MTases, is essential for RNA stability and cellular regulation. This study classifies 2′-O-MTases into 11 families, revealing their widespread distribution and distinct evolutionary trajectories. The results reveal that 2′-O-MTases expanded considerably from prokaryotes to eukaryotes, where they may be involved in regulating differentiation and development [35,36].

Among the 11 2′-O-MTase families, the Fibrillarin family exhibits remarkable evolutionary conservation both in sequence and structure, reflecting its essential role in rRNA 2′-O-methylation. Despite moderate sequence divergence (~58% similarity across kingdoms), Fibrillarin maintains high structural similarity (TM-score > 0.8), suggesting strong structural robustness. This robustness likely reflects strong purifying selection that preserves the conserved structural fold critical for its catalytic function, while still allowing for peripheral sequence variation. Although gene duplication is typically associated with functional divergence, Fibrillarin homologs—such as FBLL1 in mammals or fbl-1/fbl-2 in planarians—exhibit divergent tissue-specific expression patterns while retaining highly conserved MTase domains [18,36]. This suggests that functional diversification occurred primarily through regulatory divergence and changes in non-catalytic regions (e.g., the GAR domain), rather than alterations in the catalytic core. Therefore, strong purifying selection on the MTase domain coexists with subfunctionalization, enabling specialization without compromising essential enzymatic activity.

SpoU-methylase and FtsJ modify multiple substrates through distinct strategies. Specifically, SpoU-methylase incorporates additional domains (such as a SpoU-sub-bind-like module for rRNA targeting), whereas FtsJ becomes diversified through structural adaptations. For the FtsJ family proteins, phylogenetic analysis categorizes them into three subclasses specified for mRNA-cap, rRNA, and tRNA, respectively, each characterized by unique catalytic pocket sizes. Among them, mRNA-cap modifiers have the largest pocket for cap accommodation, rRNA modifiers maintain intermediate sizes for complex folding, and tRNA modifiers feature compact pockets for anticodon loop modification. These structural variations highlight the role of flexibility in the functional diversification of the FtsJ family, driven by evolutionary pressures to fine-tune substrate specificity.

Spatial structural analyses reveal that the catalytic domains of some 2′-O-MTases exhibit remarkable structural similarity during evolution, suggesting a potential origin from a common ancestral fold [37]. Despite sequence divergence obscuring direct phylogenetic relationships, the slower-evolving structural features retain detectable evolutionary signatures, as evidenced by the four conserved structural classes observed, which hint at a diversified origin for early 2′-O-MTase modification systems. However, the possibility of functional convergence cannot be excluded—similar structures may have arisen independently due to convergent evolution under shared catalytic geometric constraints. Future studies could resolve the homology versus convergence problem by dissecting conserved motifs (e.g., substrate-binding pockets), thereby elucidating how RNA modification tools from the LUCA era have evolved to underpin modern biological complexity.

Collectively, this study uncovers the ancient origin of diverse 2′-O-MTases, establishing a framework for understanding how RNA modification systems have become diversified to regulate biological complexity across different life domains.

4. Materials and Methods

4.1. Identification of 2′-O-MTase-Conserved, Methyltransferase-Associated Domains

To systematically identify RNA 2′-O-MTase-conserved, methyltransferase-associated domains, we conducted a comprehensive literature search using the keywords 'RNA 2′-O-methylation', 'Nm methylation', and 'RNA 2′-O-methyltransferase', focusing on enzymes validated by in vivo, in vitro, or knockdown experiments (Table S1). We then performed hmmscan analysis (HMMER v3.1b2) [38] on these known 2′-O-MTases with stringent parameters (-E 0.01 --domE 0.01 --incE 0.01) to identify conserved MTase domains.

4.2. Identification of 2′-O-MTases Across the Tree of Life

To analyze the distribution of 2′-O-MTases across different life domains, we selected 198 genomes from 44 bacterial [39,40], 32 archaeal [39,41], 29 fungal [42,43], 60 animal [44,45], and 33 plant species [46,47], representing well-sequenced organisms across life kingdoms (Table S2). Using previously identified 2′-O-MTase-related MTase domains, we performed a genome-wide hmmsearch (-E 0.01, --domE 0.01, --incE 0.01) to identify 2′-O-MTase family members. Sequences were manually curated to ensure completeness, with truncated sequences excluded.

Data retrieval and analysis utilized NCBI (https://www.ncbi.nlm.nih.gov/, last accessed on 20 May 2024), JGI (https://genome.jgi.doe.gov/portal/, last accessed on 28 May 2024), GTDB (https://gtdb.ecogenomic.org/, last accessed on 8 June 2024), and Phytozome (https://phytozome-next.jgi.doe.gov/, last accessed on 12 June 2024). Annotation files (GFF/BED) were used to filter isoforms, retaining the longest gene version for analysis.

4.3. Domain Architecture Pattern Analysis

Domain annotation for 2′-O-MTase homologous proteins was performed using hmmscan (-E 0.01, --domE 0.01, --incE 0.01). Domain architectures were manually verified for completeness and functional relevance. We analyzed frequently occurring (appearing >2 times) and functionally validated architectures and examined the co-occurrence and frequency of MTase domains with other domains.

4.4. Multiple-Sequence Alignment and Phylogenetic Analysis

Based on the hmmscan results, the MTase domain sequences from each of the 11 protein families were extracted. Pairwise alignments were then performed using Clustal Omega v1.2.4 (default parameters) [48] to calculate their sequence similarity.

To resolve phylogenetic relationships within the FtsJ family, we performed multiple-sequence alignment using Clustal Omega v1.2.4. The phylogenetic tree was constructed using IQ-TREE2 [49] with the optimal substitution model selected automatically by IQ-TREE2 using the '-m TEST' option (1000 bootstrap support). Experimentally validated proteins were mapped onto the phylogeny to analyze their functional distribution across FtsJ subclasses.

4.5. Ka/Ks Analysis

The protein sequences and coding sequences (CDSs) of each 2′-O-MTase domain were aligned and converted into codon-based alignments using ParaAT v2.0 [50], respectively. KaKs-Calculator 3.0 [51] was used to estimate nonsynonymous (Ka) and synonymous (Ks) substitution rates, as well as the Ka/Ks ratio for each alignment.

4.6. Structural Similarity Analysis and Clustering

To assess selective pressures on 2′-O-MTases from a structural perspective, we performed sequence and structural distance comparisons of representative protein domains. Protein structures were predicted using ColabFold (3 recycles, Amber relaxation) [52], and domains with a mean pLDDT score greater than 70 were extracted. TM scores were calculated using US-align [53]. Protein sequence distances were determined with MEGA12 (JTT model, 1500 bootstraps) [54], and structural distances were calculated as 1-TM-score. Linear regression was used to analyze the relationship between structural similarity and evolutionary distance [30], with adjusted R^2 indicating the strength of the correlation and 1/slope representing evolutionary structural tolerance (Table S7). Pairwise structural similarities (1-TM-score) were used to generate a distance matrix. Hierarchical clustering was performed with PHYLIP-Neighbor to construct NJ trees [55].

4.7. Polyphony Analysis

Polyphony is a differential geometry approach suitable for analyzing flexible protein structures. To perform the spatial structural classification of 2′-O-MTases, we prepared the structural data according to the literature [56]. We first used the polyphony_create_fasta_file.py code from the Polyphony suite to generate sequences and corresponding residue-level structural data. Then, we performed sequence alignment using

MAFFT [57] with the parameters --localpair and --maxiterate 1000 to ensure accurate alignment of the protein sequences. Finally, we used the integrated polyphony_Ete2_tree.py code, which incorporates multiple modules from the Polyphony project, including Structural_Alignment, Properties, Structure_Matrix, and Tree, to handle tasks such as structural alignment, property calculation, structural difference matrix generation, and tree construction. These steps enabled us to conduct a comprehensive structural analysis and classification of the 2′-O-MTases family.

4.8. Motif Identification

To identify conserved motifs in the FtsJ family, we used the MEME Suite under the ZOOPS model, allowing for flexible motif detection [58]. The analysis identified up to five motifs (4–10 residues wide), sorted by their aligned positions in subclass multiple-sequence alignments.

4.9. Tetrahedron KDKE Volume Calculation and Pairwise Distance Analysis

To quantify the KDKE catalytic motif volume, we performed multiple-sequence alignment of the MTase domain of FtsJ family proteins, identifying KDKE positions through alignment with experimentally validated active sites. The α-carbon (Cα) atoms of K1, D, K2, and E were extracted for structural analysis. The tetrahedral volume of the KDKE catalytic cage was calculated using the formula $V = \frac{1}{6}\left|\vec{v_1} \cdot \left(\vec{v_2} \times \vec{v_3}\right)\right|$. Here, $\vec{v_1}$ is the vector from point p_{K1} to point p_D, $\vec{v_2}$ is the vector from point p_{K1} to point p_{K2}, and $\vec{v_3}$ is the vector from point p_{K1} to point p_E. Additionally, pairwise distances between the Cα atoms of the KDKE residues were calculated using the Euclidean distance formula: $d = \sqrt{\left\{(x_2 - x_1)^2 + (y_2 - y_1)^2 + (z_2 - z_1)^2\right\}}$. Here, (x_1, y_1, z_1) and (x_2, y_2, z_2) are the coordinates of the two sites.

5. Conclusions

Based on unique MTase domains, we categorized 2′-O-MTases into 11 families based on a collection of functionally reported 2′-O-MTases. Overall, 6746 homolog proteins were identified via a domain-based search against 198 species in all living kingdoms, indicating that 2′-O-MTases have been widely distributed throughout the evolution of life. Of the 11 families of 2′-O-MTases, 8 were widely spread in both eukaryotes and prokaryotes, while the remaining ones were life domain-specific. Purifying selection has allowed for conservative evolution of the catalytic domain at both the sequence and structural levels, despite the fact that the integration of additional domains occurred frequently throughout the evolution of most 2′-O-MTase families, according to domain composition, Ka/Ks ratio, and domain structural analyses. In conclusion, our research sheds light on the evolutionary history of 2′-O-MTases, emphasizing their numerous ancient origins and functional diversity.

Supplementary Materials: The following supporting information can be downloaded at https://www.mdpi.com/article/10.3390/ijms26115260/s1.

Author Contributions: Conceptualization, S.-N.W. and Y.L.; software, L.-J.L. and Q.W.; validation, S.-N.W. and X.-X.L.; investigation, X.-X.L.; data curation, L.-J.L.; writing—original draft preparation, S.-N.W., X.-X.L. and L.-J.L.; writing—review and editing, Y.L. and Z.-Q.S.; visualization, S.-N.W. and X.-X.L.; supervision, Y.L., Z.-Q.S. and Q.W. All authors have read and agreed to the published version of the manuscript.

Funding: This research received no external funding.

Institutional Review Board Statement: Not applicable.

Informed Consent Statement: Not applicable.

Data Availability Statement: Data are contained within the article and Supplementary Materials.

Conflicts of Interest: The authors declare no conflicts of interest.

Abbreviations

The following abbreviations are used in this manuscript:

Nm	2′-O-methylation
2′-O-MTase	2′-O-methyltransferase
MTase	Methyltransferase
LUCA	Last Universal Common Ancestor
Ka	Nonsynonymous substitution rate
Ks	Synonymous substitution rate
m^6A	N6-methyladenosine
m^5C	5-methylcytosine
m^1A	N1-methyladenosine
m^7G	N7-methylguanosine
ac^4C	N4-acetylcytosine
Ψ	pseudouridine
SAM	S-adenosylmethionine
piRNA	Piwi-interacting RNA
miRNA	MicroRNA
snoRNP	Small nucleolar ribonucleoprotein

References

1. Goldberg, A.D.; Allis, C.D.; Bernstein, E. Epigenetics: A landscape takes shape. *Cell* **2007**, *128*, 635–638. [CrossRef] [PubMed]
2. Zhou, K.I.; Pecot, C.V.; Holley, C.L. 2′-O-methylation (Nm) in RNA: Progress, challenges, and future directions. *RNA* **2024**, *30*, 570–582. [CrossRef] [PubMed]
3. Khoshnevis, S.; Dreggors-Walker, R.E.; Marchand, V.; Motorin, Y.; Ghalei, H. Ribosomal RNA 2′-O-methylations regulate translation by impacting ribosome dynamics. *Proc. Natl. Acad. Sci. USA* **2022**, *119*, e2117334119. [CrossRef] [PubMed]
4. Zhang, M.; Lu, Z. tRNA modifications: Greasing the wheels of translation and beyond. *RNA Biol.* **2025**, *22*, 1–25. [CrossRef]
5. Li, J.; Yang, Z.; Yu, B.; Liu, J.; Chen, X. Methylation protects miRNAs and siRNAs from a 3′-end uridylation activity in *Arabidopsis*. *Curr. Biol.* **2005**, *15*, 1501–1507. [CrossRef]
6. Ramanathan, A.; Robb, G.B.; Chan, S.H. mRNA capping: Biological functions and applications. *Nucleic Acids Res.* **2016**, *44*, 7511–7526. [CrossRef]
7. Ge, J.; Liu, H.; Yu, Y.T. Regulation of pre-mRNA splicing in Xenopus oocytes by targeted 2′-O-methylation. *RNA* **2010**, *16*, 1078–1085. [CrossRef]
8. Tang, Y.; Wu, Y.; Wang, S.; Lu, X.; Gu, X.; Li, Y.; Yang, F.; Xu, R.; Wang, T.; Jiao, Z.; et al. An integrative platform for detection of RNA 2′-O-methylation reveals its broad distribution on mRNA. *Cell Rep. Methods* **2024**, *4*, 100721. [CrossRef]
9. Rodriguez-Corona, U.; Sobol, M.; Rodriguez-Zapata, L.C.; Hozak, P.; Castano, E. Fibrillarin from Archaea to human. *Biol. Cell* **2015**, *107*, 159–174. [CrossRef]
10. Pintard, L.; Lecointe, F.; Bujnicki, J.M.; Bonnerot, C.; Grosjean, H.; Lapeyre, B. Trm7p catalyses the formation of two 2′-O-methylriboses in yeast tRNA anticodon loop. *EMBO J.* **2002**, *21*, 1811–1820. [CrossRef]
11. Drazkowska, K.; Tomecki, R.; Warminski, M.; Baran, N.; Cysewski, D.; Depaix, A.; Kasprzyk, R.; Kowalska, J.; Jemielity, J.; Sikorski, P.J. 2′-O-Methylation of the second transcribed nucleotide within the mRNA 5′ cap impacts the protein production level in a cell-specific manner and contributes to RNA immune evasion. *Nucleic Acids Res.* **2022**, *50*, 9051–9071. [CrossRef] [PubMed]
12. Devarkar, S.C.; Wang, C.; Miller, M.T.; Ramanathan, A.; Jiang, F.; Khan, A.G.; Patel, S.S.; Marcotrigiano, J. Structural basis for m7G recognition and 2′-O-methyl discrimination in capped RNAs by the innate immune receptor RIG-I. *Proc. Natl. Acad. Sci. USA* **2016**, *113*, 596–601. [CrossRef]
13. Decombe, A.; El Kazzi, P.; Decroly, E. Interplay of RNA 2′-O-methylations with viral replication. *Curr. Opin. Virol.* **2023**, *59*, 101302. [CrossRef] [PubMed]
14. Park, G.J.; Osinski, A.; Hernandez, G.; Eitson, J.L.; Majumdar, A.; Tonelli, M.; Henzler-Wildman, K.; Pawłowski, K.; Chen, Z.; Li, Y.; et al. The mechanism of RNA capping by SARS-CoV-2. *Nature* **2022**, *609*, 793–800. [CrossRef]

15. Deng, J.; Yang, S.; Li, Y.; Tan, X.; Liu, J.; Yu, Y.; Ding, Q.; Fan, C.; Wang, H.; Chen, X.; et al. Natural evidence of coronaviral 2′-O-methyltransferase activity affecting viral pathogenesis via improved substrate RNA binding. *Signal Transduct. Target. Ther.* **2024**, *9*, 140. [CrossRef] [PubMed]
16. Bergant, V.; Yamada, S.; Grass, V.; Tsukamoto, Y.; Lavacca, T.; Krey, K.; Mühlhofer, M.T.; Wittmann, S.; Ensser, A.; Herrmann, A.; et al. Attenuation of SARS-CoV-2 replication and associated inflammation by concomitant targeting of viral and host cap 2′-O-ribose methyltransferases. *EMBO J.* **2022**, *41*, e111608. [CrossRef]
17. Pereira-Santana, A.; Gamboa-Tuz, S.D.; Zhao, T.; Schranz, M.E.; Vinuesa, P.; Bayona, A.; Rodríguez-Zapata, L.C.; Castano, E. Fibrillarin evolution through the Tree of Life: Comparative genomics and microsynteny network analyses provide new insights into the evolutionary history of Fibrillarin. *PLoS Comput. Biol.* **2020**, *16*, e1008318. [CrossRef]
18. Chen, J.; Li, Y.; Wang, Y.; Wang, H.; Yang, J.; Pan, X.; Zhao, Y.; Xu, H.; Jiang, P.; Qian, P.; et al. Fibrillarin homologs regulate translation in divergent cell lineages during planarian homeostasis and regeneration. *EMBO J.* **2024**, *43*, 6591–6625. [CrossRef]
19. Hajieghrari, B.; Niazi, A. Phylogenetic and Evolutionary Analysis of Plant Small RNA 2′-O-Methyltransferase (HEN1) Protein Family. *J. Mol. Evol.* **2023**, *91*, 424–440. [CrossRef]
20. Tkaczuk, K.L.; Dunin-Horkawicz, S.; Purta, E.; Bujnicki, J.M. Structural and evolutionary bioinformatics of the SPOUT superfamily of methyltransferases. *BMC Bioinform.* **2007**, *8*, 73. [CrossRef]
21. Li, J.; Wang, Y.N.; Xu, B.S.; Liu, Y.P.; Zhou, M.; Long, T.; Li, H.; Dong, H.; Nie, Y.; Chen, P.R.; et al. Intellectual disability-associated gene ftsj1 is responsible for 2′-O-methylation of specific tRNAs. *EMBO Rep.* **2020**, *21*, e50095. [CrossRef] [PubMed]
22. Ringeard, M.; Marchand, V.; Decroly, E.; Motorin, Y.; Bennasser, Y. FTSJ3 is an RNA 2′-O-methyltransferase recruited by HIV to avoid innate immune sensing. *Nature* **2019**, *565*, 500–504. [CrossRef] [PubMed]
23. Wilkinson, M.L.; Crary, S.M.; Jackman, J.E.; Grayhack, E.J.; Phizicky, E.M. The 2′-O-methyltransferase responsible for modification of yeast tRNA at position 4. *RNA* **2007**, *13*, 404–413. [CrossRef] [PubMed]
24. Kimura, S.; Suzuki, T. Fine-tuning of the ribosomal decoding center by conserved methyl-modifications in the *Escherichia coli* 16S rRNA. *Nucleic Acids Res.* **2010**, *38*, 1341–1352. [CrossRef]
25. Persson, B.C.; Jäger, G.; Gustafsson, C. The *spoU* gene of *Escherichia coli*, the fourth gene of the *spoT* operon, is essential for tRNA (Gm18) 2′-O-methyltransferase activity. *Nucleic Acids Res.* **1997**, *25*, 4093–4097. [CrossRef]
26. Cavaillé, J.; Chetouani, F.; Bachellerie, J.P. The yeast *Saccharomyces cerevisiae* YDL112w ORF encodes the putative 2′-O-ribose methyltransferase catalyzing the formation of Gm18 in tRNAs. *RNA* **1999**, *5*, 66–81. [CrossRef]
27. Purta, E.; van Vliet, F.; Tkaczuk, K.L.; Dunin-Horkawicz, S.; Mori, H.; Droogmans, L.; Bujnicki, J.M. The *yfhQ* gene of *Escherichia coli* encodes a tRNA:Cm32/Um32 methyltransferase. *BMC Mol. Biol.* **2006**, *7*, 23. [CrossRef]
28. Benítez-Páez, A.; Villarroya, M.; Douthwaite, S.; Gabaldón, T.; Armengod, M.E. YibK is the 2′-O-methyltransferase TrmL that modifies the wobble nucleotide in *Escherichia coli* tRNA(Leu) isoacceptors. *RNA* **2010**, *16*, 2131–2143. [CrossRef]
29. Kressler, D.; Rojo, M.; Linder, P.; Cruz, J. Spb1p is a putative methyltransferase required for 60S ribosomal subunit biogenesis in *Saccharomyces cerevisiae*. *Nucleic Acids Res.* **1999**, *27*, 4598–4608. [CrossRef]
30. Illergård, K.; Ardell, D.H.; Elofsson, A. Structure is three to ten times more conserved than sequence—A study of structural response in protein cores. *Proteins* **2009**, *77*, 499–508. [CrossRef]
31. Egloff, M.P.; Benarroch, D.; Selisko, B.; Romette, J.L.; Canard, B. An RNA cap (nucleoside-2′-O-)-methyltransferase in the flavivirus RNA polymerase NS5: Crystal structure and functional characterization. *EMBO J.* **2002**, *21*, 2757–2768. [CrossRef] [PubMed]
32. Smietanski, M.; Werner, M.; Purta, E.; Kaminska, K.H.; Stepinski, J.; Darzynkiewicz, E.; Nowotny, M.; Bujnicki, J.M. Structural analysis of human 2′-O-ribose methyltransferases involved in mRNA cap structure formation. *Nat. Commun.* **2014**, *5*, 3004. [CrossRef]
33. Punekar, A.S.; Shepherd, T.R.; Liljeruhm, J.; Forster, A.C.; Selmer, M. Crystal structure of RlmM, the 2′O-ribose methyltransferase for C2498 of *Escherichia coli* 23S rRNA. *Nucleic Acids Res.* **2012**, *40*, 10507–10520. [CrossRef]
34. Werner, M.; Purta, E.; Kaminska, K.H.; Cymerman, I.A.; Campbell, D.A.; Mittra, B.; Zamudio, J.R.; Sturm, N.R.; Jaworski, J.; Bujnicki, J.M. 2′-O-ribose methylation of cap2 in human: Function and evolution in a horizontally mobile family. *Nucleic Acids Res.* **2011**, *39*, 4756–4768. [CrossRef]
35. Häfner, S.J.; Jansson, M.D.; Altinel, K.; Andersen, K.L.; Abay-Nørgaard, Z.; Ménard, P.; Fontenas, M.; Sørensen, D.M.; Gay, D.M.; Arendrup, F.S.; et al. Ribosomal RNA 2′-O-methylation dynamics impact cell fate decisions. *Dev. Cell* **2023**, *58*, 1593–1609.e9. [CrossRef] [PubMed]
36. Zhang, D.; Li, B.; Xu, H.; Li, J.; Ma, C.; Ge, W.; Lu, C.; Cao, X. Identification of FBLL1 as a neuron-specific RNA 2′-O-methyltransferase mediating neuronal differentiation. *Proc. Natl. Acad. Sci. USA* **2024**, *121*, e2406961121. [CrossRef] [PubMed]
37. Vaysset, H.; Meers, C.; Cury, J.; Bernheim, A.; Sternberg, S.H. Evolutionary origins of archaeal and eukaryotic RNA-guided RNA modification in bacterial IS110 transposons. *Nat. Microbiol.* **2025**, *10*, 20–27. [CrossRef]
38. Eddy, S.R. Accelerated Profile HMM Searches. *PLoS Comput. Biol.* **2011**, *7*, e1002195. [CrossRef]

39. Parks, D.H.; Chuvochina, M.; Rinke, C.; Mussig, A.J.; Chaumeil, P.A.; Hugenholtz, P. GTDB: An ongoing census of bacterial and archaeal diversity through a phylogenetically consistent, rank normalized and complete genome-based taxonomy. *Nucleic Acids Res.* **2022**, *50*, D785–D794. [CrossRef]
40. Parks, D.H.; Chuvochina, M.; Waite, D.W.; Rinke, C.; Skarshewski, A.; Chaumeil, P.A.; Hugenholtz, P. A standardized bacterial taxonomy based on genome phylogeny substantially revises the tree of life. *Nat. Biotechnol.* **2018**, *36*, 996–1004. [CrossRef]
41. Ghaly, T.M.; Tetu, S.G.; Penesyan, A.; Qi, Q.; Rajabal, V.; Gillings, M.R. Discovery of integrons in Archaea: Platforms for cross-domain gene transfer. *Sci. Adv.* **2022**, *8*, eabq6376. [CrossRef] [PubMed]
42. Naranjo-Ortiz, M.A.; Gabaldón, T. Fungal evolution: Diversity, taxonomy and phylogeny of the Fungi. *Biol. Rev. Camb. Philos. Soc.* **2019**, *94*, 2101–2137. [CrossRef] [PubMed]
43. Spatafora, J.W.; Aime, M.C.; Grigoriev, I.V.; Martin, F.; Stajich, J.E.; Blackwell, M. The Fungal Tree of Life: From Molecular Systematics to Genome-Scale Phylogenies. *Microbiol. Spectr.* **2017**, *5*, 10.1128. [CrossRef] [PubMed]
44. Simakov, O.; Kawashima, T.; Marlétaz, F.; Jenkins, J.; Koyanagi, R.; Mitros, T.; Hisata, K.; Bredeson, J.; Shoguchi, E.; Gyoja, F.; et al. Hemichordate genomes and deuterostome origins. *Nature* **2015**, *527*, 459–465. [CrossRef]
45. Telford, M.J.; Budd, G.E.; Philippe, H. Phylogenomic Insights into Animal Evolution. *Curr. Biol.* **2015**, *25*, R876–R887. [CrossRef]
46. Zhang, Z.; Ma, X.; Liu, Y.; Yang, L.; Shi, X.; Wang, H.; Diao, R.; Zhong, B. Origin and evolution of green plants in the light of key evolutionary events. *J. Integr. Plant Biol.* **2022**, *64*, 516–535. [CrossRef]
47. Li, S.X.; Liu, Y.; Zhang, Y.M.; Chen, J.Q.; Shao, Z.Q. Convergent reduction of immune receptor repertoires during plant adaptation to diverse special lifestyles and habitats. *Nat. Plants* **2025**, *11*, 248–262. [CrossRef]
48. Sievers, F.; Wilm, A.; Dineen, D.; Gibson, T.J.; Karplus, K.; Li, W.; Lopez, R.; McWilliam, H.; Remmert, M.; Söding, J.; et al. Fast, scalable generation of high-quality protein multiple sequence alignments using Clustal Omega. *Mol. Syst. Biol.* **2011**, *7*, 539. [CrossRef]
49. Minh, B.Q.; Schmidt, H.A.; Chernomor, O.; Schrempf, D.; Woodhams, M.D.; von Haeseler, A.; Lanfear, R. IQ-TREE 2: New Models and Efficient Methods for Phylogenetic Inference in the Genomic Era. *Mol. Biol. Evol.* **2020**, *37*, 1530–1534. [CrossRef]
50. Zhang, Z.; Xiao, J.; Wu, J.; Zhang, H.; Liu, G.; Wang, X.; Dai, L. ParaAT: A parallel tool for constructing multiple protein-coding DNA alignments. *Biochem. Biophys. Res. Commun.* **2012**, *419*, 779–781. [CrossRef]
51. Zhang, Z. KaKs_Calculator 3.0: Calculating Selective Pressure on Coding and Non-coding Sequences. *Genom. Proteom. Bioinform.* **2022**, *20*, 536–540. [CrossRef] [PubMed]
52. Mirdita, M.; Schütze, K.; Moriwaki, Y.; Heo, L.; Ovchinnikov, S.; Steinegger, M. ColabFold: Making protein folding accessible to all. *Nat. Methods* **2022**, *19*, 679–682. [CrossRef] [PubMed]
53. Zhang, C.; Shine, M.; Pyle, A.M.; Zhang, Y. US-align: Universal structure alignments of proteins, nucleic acids, and macromolecular complexes. *Nat. Methods* **2022**, *19*, 1109–1115. [CrossRef]
54. Kumar, S.; Stecher, G.; Suleski, M.; Sanderford, M.; Sharma, S.; Tamura, K. MEGA12: Molecular Evolutionary Genetic Analysis version 12 for adaptive and green computing. *Mol. Biol. Evol.* **2024**, *41*, msae263. [CrossRef]
55. Saitou, N.; Nei, M. The neighbor-joining method: A new method for reconstructing phylogenetic trees. *Mol. Biol. Evol.* **1987**, *4*, 406–425. [CrossRef] [PubMed]
56. Pitt, W.R.; Montalvão, R.W.; Blundell, T.L. Polyphony: Superposition independent methods for ensemble-based drug discovery. *BMC Bioinform.* **2014**, *15*, 324. [CrossRef]
57. Katoh, K.; Standley, D.M. MAFFT multiple sequence alignment software version 7: Improvements in performance and usability. *Mol. Biol. Evol.* **2013**, *30*, 772–780. [CrossRef]
58. Bailey, T.L.; Elkan, C. Fitting a mixture model by expectation maximization to discover motifs in biopolymers. *Proc. Int. Conf. Intell. Syst. Mol. Biol.* **1994**, *2*, 28–36.

MDPI AG
Grosspeteranlage 5
4052 Basel
Switzerland
Tel.: +41 61 683 77 34

International Journal of Molecular Sciences Editorial Office
E-mail: ijms@mdpi.com
www.mdpi.com/journal/ijms

www.ingramcontent.com/pod-product-compliance
Lightning Source LLC
LaVergne TN
LVHW070444061126
851858LV00011B/160

* 9 7 8 3 7 2 5 8 8 0 7 5 1 *